BUGGING
CANCER

BUGGING CANCER
DARING TO DREAM

Authors (in order of chapters written):
Indrani Mondal, Ranjita Chattopadhyay, Ranajoy Ganguli,
Bakul Banerjee, Debkumar Chatterjee, Debabrata
Chakravarti, Krishna Chakrabarty, Ananda Chakrabarty

LOGOS PRESS

BUGGING CANCER
DARING TO DREAM

Ananda Chakrabarty and the Chicago Onco-Group
Authors (in order of chapters written): Indrani Mondal, Ranjita Chattopadhyay, Ranajoy Ganguli, Bakul Banerjee, Debkumar Chatterjee, Debabrata Chakravarti, Krishna Chakrabarty, Ananda Chakrabarty

Published in The United States of America
by
Logos Press®, Washington, DC
www.Logos-Press.com
info@Logos-Press.com

10 9 8 7 6 5 4 3 2 1

ISBN-13
Hardcover: 978-1-934899-37-3
Softcover: 978-1-934899-36-6

Contents

Preface

For centuries mankind has been intimidated by cancer, and research has been going on for many decades to address or attack the disease meaningfully so as to eradicate it. But the causes of cancer are so diverse in nature that it is extremely difficult to predict or to control the conditions, let alone fully cure it. Hundreds of mutations in many genes have been linked to different cancers. Some are described as driver mutations, while other mutations are passenger mutations. Nonetheless, significant strides have been made in developing therapeutics that target these driver events, thereby controlling the growth of cancer. There has also been extensive research on cancer with an attempt to use live bacteria, often called *bugs*, and/or bacterial proteins of a certain nature to attack malignant tumor cells and that is going to be the central theme of this book and hence the title; along with the associated legal, medical and moral issues encompassing the theme.

The first thing the authors would like to acknowledge is that this book is a work of fiction, with embedded bits of reality. The story is based on a real bacterial peptide called p28. A peptide is part of a protein. This peptide is called p28 because it contains 28 amino acid building blocks. It has preferential entry into cancer cells, cancer killing ability and even cancer preventive activity. Phase I clinical trials have shown that p28 has demonstrated no toxicity in animal studies or human cancer patients, and significant beneficial effects in several stage IV cancer patients for whom no conventional drugs were working, and who had a life expectancy of less than 6 months. Intravenous injections of p28 in these 15 stage IV cancer patients, while demonstrating very few side effects, showed stabilization of tumor growth in several patients, partial regression in three patients and complete regression in one other, demonstrating that p28 has cancer killing activity in patients where the cancers were no longer responding to traditional drugs. Three of these patients were alive after more than 2 to 3 years of the start of the trial, one completely disease free.

So what is the point, and why this book? This book extends the p28 discoveries in a fictional way to describe the cancer-fighting power of an imaginary bacterial protein termed neelazin. This protein is meant to resemble p28 and its parent protein azurin. It was developed by Sam, the scientist in our story, in an Arctic island. Prior to developing neelazin, Sam had worked in Milanburg, a fictional suburb of Chicago and discovered another bacterial protein called rhabdosin, also fictional. The clinical trials were conducted by Brian, the gene therapist, and Sam, the inventor, in a remote island called Bluefrostland, also fictional, somewhere north of the Arctic circle; an island in which day to day life is controlled by scientific facts and where government encourages risk-taking to apply human imagination to solve major problems. We concocted this scenario, where our anticancer protein works in curing severely ill patients. Can it be real by any stretch of the imagination? We believe it may. We have included a few scientific references and a glossary at the end of the book. We will let the readers draw their own conclusions and be eager to hear if they are daring to dream that one day cancer can be bugged and stopped.

Why bacteria? The question really should be why not? We all know how beneficial our intestinal microbiota are to our own survival. Therefore, the thought of a bacterium helping cure a deadly disease like cancer is a dream worth having and a dream worth daring. Why use a fictional bacterial protein? Why would bacteria make such a protein fight cancer? What grudge do bacteria have against cancer? It has actually been known for over 100 years that bacteria don't like cancer when they infect humans. Many people are working to develop genetically engineered bacteria to keep cancer in check. Unfortunately, using live bacteria to fight cancer has met with challenges of strong immune response from the patients, sometimes making them more sick.

The authors believe that many pathogenic bacteria, which live in the human body for months or even years by forming a thin film of growth called a biofilm, consider the human body as their home and try to prevent intruders like cancers, viruses and parasites that can cause harm to their habitat, the human body, and so they have used their evolutionary wisdom to develop an intelligently-designed protein weapon such as neelazin. As opposed to small molecule drugs that are favored today by the pharmaceutical industry, we point to the potential value of bacterial protein/peptide drugs with both cancer therapeutic and preventive activity to be expressed from the human genome to guard against cancer. This is an unproven and absolutely hypothetical scenario that merits some debate, as discussed in Chapter 22.

There is one more idea the book is trying to convey.

Cancer is basically a disease involving human gene mutations, often caused by mutagens and carcinogens in our environment or simply due to accumulation of mutations in our genes as we age. Some of these mutations can be passed on through generations, as depicted for Erin, our heroine, in chapter 1. The situation is complicated because of patenting of the testing methods for such mutations. This has allowed the book to deal with certain aspects of patent controversies and court cases. The patent laws are in the US Constitution, and the first patent was granted on July 31, 1790, signed by President George Washington and Secretary of State Thomas Jefferson, demonstrating the importance of the patenting process our forefathers envisioned. Indeed, President Abraham Lincoln was the holder of a patent. The book thus raises a hypothetical scenario of introducing the bacterial gene neelazin in the human genome to act as a guard against diseases such as cancer, but also creating a de facto genetically-engineered human being. The societal concerns of attempting to manipulate human genome, including possibly inserting a bacterial gene, even for a worthy cause, and patent claims of such genetic engineering techniques to protect human health through surveillance against cancers, will likely be the subject of intense debates.

Finally, this book has been written by multiple authors. Even though several of us collaborated in writing some chapters, each chapter has a primary author. Erin's character has been developed by Indrani Mondal. Sam was conceived by Ranajoy Ganguli. Nick is the creation of Ranjita Chattopadhyay. Bakul Banerjee penned George's story. Brian has arisen from the imagination of Debkumar Chatterjee. Debabrata Chakravarti and Debkumar Chatterjee, both scientists, are responsible for the scientific content and conversations in various chapters. Krishna Chakrabarty assembled the different chapters into one cohesive document, and wrote the last chapter. Finally, writing this book was Ananda Chakrabarty's idea. He and his colleagues conducted the original research on azurin and p28 (see the Reference section). He also wrote the Epilogue. The authors would love to know what the readers think.

one
Erin's story

After the train ride, it was barely a fifteen minute walk to Aurum University library in Milanburg, or a ten minute bus ride. It was a gorgeous spring day. A little on the cool side, no doubt, but that made the bright sunshine even more welcome. Erin decided to walk. The tall trees that had stood bare and bereft over the long winter months had smears of fresh green from sprouting new leaves. Daffodils were nodding on doorsteps and lilies, almost ready to bloom, along sidewalks. Erin's heart throbbed. She remembered her colleague Preeti telling her that in her culture when parents died, they became stars that watch over their children from above. Erin would much rather prefer them to be flowers, such as these early spring blooms, lending their delightful color close to earth, so that she could touch and feel them and be assured of their continued presence. Yes, today there was life and hope everywhere. Everywhere there was expectancy and awakening. Erin walked rather slowly, not because she was enjoying this new spring day. It was because anything beautiful reminded her of her mom and how beautiful things are short lived. Erin felt cold even in the sunshine. She vividly remembered her mom and her agony. Hard as it was even now, Erin would have to accept that one way of life was definitely over. "Steady there, Bonita!" she chided herself, using the penname she signed with on her poetry blog. It was also her mother's endearment for her as a school girl when she felt sad. Unconsciously she fingered the necklace she always wore. It was a pretty silver chain with 'Bonita' engraved on the pendent, which her mom had given her on her sixteenth birthday.

Six years ago Nancy Goldberg, Erin's mother, had succumbed to breast cancer. Erin was barely twenty then. After high school she was taking library science courses at the local community college and working as a grant writer for a non-profit organization to make ends meet, as Nancy was a single parent. Mother and daughter had a small apartment uptown, where the lack of glamour was compensated by warm neighborly support. Nancy had been an assistant to Dr. Zen Lee, a general physician, who had a small downtown office. Erin had always thought it ironic how her mom had spent hours documenting other patients' illnesses but by the time Dr. Zen Lee had forced her to take her own spells of weakness and physical checkups more seriously, Nancy's disease had progressed irreversibly. Erin's mother had not been able to celebrate her fortieth birthday with her sprightly young daughter. After a few short hectic months of fighting breast cancer, going back and forth from the hospital for radiation and chemo, as other invasive procedures were frantically being discussed by the doctors, Nancy had died unceremoniously in her hospital room. Erin would never forget that fateful day which was also Nancy's birthday. On her lunch break Erin had been on her way to visit Nancy with red roses and a musical card to wish her mom happy fortieth in style. The roses had never reached Nancy, only her hearse. Erin still remembered how ridiculously out of place the bright red roses had looked amidst the other white flowers. Almost as ridiculous and totally unfair as her mom's passing away and being carried to a funeral home on her fortieth birthday, without even bidding her only child a real goodbye.

Afterwards, Dr. Zen Lee, the industrious and energetic doctor, who was also a kind man, offered Erin a part time position in his office. Erin had just applied for her present library job at that time and was still waiting to hear from them. She accepted Dr. Lee's offer, for in some strange way she sensed that being in the same office where her mom had spent the greater part of her days would make her feel closer to her lost parent, even though it might be for a few short hours each week. The very next month she had also been hired to the Aurum University library administration team for their reputed reference section. From then on Erin's real full time job of juggling work schedules and postponing dreams had started. Just before Nancy fell seriously ill, Erin had been planning to apply to the university undergrad program with an education major and language and arts minor. Overnight her hopes of higher studies had to be put on hold for the basic demands of day-to-day living and helping pay off her mom's hefty medical bills. Her colleague Preeti's gift of the *Gita*, an Indian fable cum scripture book, had given Erin a different take on life and death. The soul never dies, she'd read, such an absolutely consoling idea, for a sorely grieving Erin, hard as it was to accept!

Unconsciously tears welled up and rolled down Erin's cheeks under her dark glasses. A few passersby glanced in her direction. But then a tall, thin young woman in casual chic dress jeans, navy sweater and brown leather jacket, highlighting her smartly layered auburn hair streaked with honey gold, around a sharp boned, lightly tanned face was bound to attract attention. On impulse Erin turned into a side street. This was a longer route to the library and she knew she would be late for work. She had recently been promoted to a full time job in the reference section. From her smart phone she quickly sent an email to her colleague Miguel to take care of the first two morning meetings. The memory of her mother's death was just starting to heal for Erin, but she never knew where and when a sight, sound or smell would unsettle the thin growing scab of memory leaving the scar bare and raw again.

After her mom passed away, Erin's high school sweetheart Derek and his family had started pressuring her to get married. He just couldn't understand Erin's need for time to figure out where she stood and where she needed to go. They had finally decided to break up and Erin had been left with a backlash of bitterness and guilt.

One evening around this time, at a library talk session during *Diverse Cultures and Rainbow of Beliefs* week, Erin had shared her understanding of the Indian spiritual text *Gita*, in a contemporary perspective. She was surprised to see that her talk session was well attended. In fact, afterwards, Dr. Sam Roy, a medical research scholar, complimented her and asked her out for coffee. She had told him, after her mom passed away suddenly from breast cancer, her friend Preeti had first read parts of the *Gita* at Nancy's funeral. Erin had loved it and so Preeti had presented her a beautifully illustrated copy of the *Gita* with English translations. Sam told her he was impressed by her open mind and ability to explore new ways of understanding.

Erin's mom had a younger sister, Angela, who had just moved to Erin's town from the east. She was unable to attend Nancy's funeral. She had married a rich banker and had a nice home in a north suburb and soon became Erin's surrogate mom, treating her to home cooked meals and cozy family weekends. Angela's daughter Kathy, who was a few months older than Erin, also became her close friend. Kathy, a pharmacist, worked in a downtown drug store and Erin and Kathy would often meet for a quick weekday lunch or a weekend happy hour at one of the downtown bars. Erin told Kathy about Sam and how he seemed fascinated by her approach to his culture and beliefs. Kathy had laughed and poked her in the ribs, "Oh come on Erin! Of course he is fascinated! But with something more than just your approach, silly!"

Melodramatic as it seemed in retrospect, less than 16 months after Angela returned to town, her routine mammogram revealed a hidden tumor in her right breast, and a biopsy showed it to be malignant. Erin knew her grandparents had died young, and Dr. Lee, who was their family primary, told Erin that when he had treated Nancy, he had found out that both her grandparents had succumbed to cancer. So their family had a long history of carrying mutations that may have increased the risk of cancer. Angela went through prophylactic surgery and double mastectomy. Both of her breasts—even the healthy one—and her ovaries were removed as a precaution. Her cancer was gone now, but she was put on intense surveillance with bone scans and lab work done on her every six months.

Erin was scared, but after educating herself in recent medical research, she found out that breast cancer has the highest survival rate if detected early enough. Still the sense of doom seemed to be her constant shadow. It seemed like wherever she found refuge, a dark cloud lurked round the corner. Around this time Kathy got a good job offer in the west and moved out. Erin's work kept her busy, no doubt, but inside she felt more alone than ever.

Erin's love of language and literature had intensified because of her constant exposure to a wealth of written and online material at the library where she worked. Every Thursday evening she attended the Creative Literary Circle, "Beyond Words." She was pleasantly surprised to run into Sam again at one of those sessions. In answer to her rather blunt question as to why he liked poetry even though he was a medical researcher, he answered that literature and writing were like his light and air, while his research work was his food and water. He needed them all for a complete life. Erin had laughed it off, but thinking about it later she kind of saw what he meant.

"Quite a guy!" she mused, "and he doesn't even give off the onion-garlicky Indian curry smell!" They got into the practice of meeting for a quick bite over a cup of coffee at the local café and often Erin would try out her aunt Angela's recipes on him. He always complimented her on her culinary concoctions, for they were another aspect of her creative appeal, as he put it. But Erin could never be sure if he especially liked her, or if his compliments were just a result of his naturally polite and affable personality.

As they munched and talked in the café, Sam and Erin had

gotten into the habit of telling each other interesting things about their day. Erin was always brainstorming how to raise money for the library to acquire new research materials and make them more accessible for researchers. Sam told her some of the most unbelievable stories of how his experiments might one day help people all over the world. In fact one evening he actually left her crispy snacks aside and told her some fantastic results he was having with experiments on mice determining causes and potential treatments for cancer.

Sam had taken the lead in the conversation. He was very good at explaining complex scientific facts to people with no scientific background. "Remember Erin, I was telling you about the work that this graduate student was doing in the lab?" He had been really excited!

"Yea, yea, I do, it is some kind of protein that does something, right?"

"Hmm, hmm. OK, Listen very carefully to what I tell you. This is going to be the best research paper we have published from the lab. Yes it is about a protein, but not the kind of protein you eat from a protein bar, but one of a kind made by a bug."

"A bug, oh my, oh my! Are you planning to put them in a protein bar and make a lot of money?" Erin had made a face.

"Would you listen and let me finish?"

"Sorry, your honor, please continue."

"This protein that I am talking about is made by a bug, a very bad bug that not only lives in our mouth but can sometimes invade the brain. It has a fancy name: *Rhabdosis pulmoneria*. Like us, these bugs or bacteria have a proper name and a surname! But, all jokes aside, remember even the bad bugs can also do good."

"After serving time!"

"Oh, come on! Listen," Sam looked very serious as he continued. "This is what my student found. Remember we used human cells and added our bacterial protein we named rhabdosin, and all the cancer cells died, but normal cells were not affected much?

This is a very exciting result by itself. Why? Because, a problem with chemotherapy is that along with bad cancer cells, our normal cells also die, and that's not good, right?"

Erin agreed taking a long sip of her tea.

"Now we feel that we have a protein that can kill only cancer cells, without affecting normal cells. So I asked my student to test it."

"Test it where? Is it like doing a test drive of a new car?"

"Kind of similar. We have a good feeling that it should work in animals but until you do it, you never know. So I like your analogy in that sense. What we did was a great experiment. We prepared a lot of rhabdosin protein. Then we took two groups of ten mice. Both groups were implanted with breast cancer cells under their skin. Then, we allowed the breast cancer cells to grow uncontrollably under the skin, you could even see the tumors with naked eyes. It was just beautiful!"

"Beautiful? You are sick!"

"Wait, let me finish. Remember the mice that had the tumors? We injected one group with rhabdosin protein three times a week for 4 weeks, and waited. We went to the mouse room like every 5 hours. After a week or ten days, you would not believe, the mice that received rhabdosin showed significant shrinkage of their tumors, while the cancer kept growing in the other group that did not receive rhabdosin."

"Are you kidding me?" Erin had looked up from her plate of samosas with wide-open eyes.

"No. There is more! For one group of mice, we gave them rhabdosin at the same time as when we also planted breast cancer cells in them. You are a very smart lady—tell me what we found?"

"Not to disappoint you but I say, there was no tumor, at least not the big ones!" Erin had quipped.

"You are a genius."

"No, you are. Are we still doing dinner with my friends tonight?"

"Of course, if you listen to my final statements, dinner is on me."

"OK, free dinner date, please continue."

"OK, we put cancer cells under the skin, and rhabdosin kills them, no big deal. It's a proof of principle, but breast cancer does not occur under the skin, it develops in your, I mean, you know what I mean, in real breasts. So we did another experiment. In this case we took genetically engineered mice that have mutations in two genes that make them develop cancers. This is an animal model to study human lung/breast cancer and drug efficacy."

"By changing mutant genes you can induce cancer in animals?"

"Of course. But some people are opposed to this kind of approach. I respect their opinion, but on the other hand how do you study complicated human diseases such as cancer. For example, we all hear about certain gene mutations that cause breast cancer in humans. How do we really know whether gene mutations actually can lead to cancer if we don't demonstrate it. Do you know what I am saying?"

"What are you saying, Sam?"

"We generate the mutations in mice and wait to see if they indeed get cancer or not. If they do then there is a great chance that these mutations probably lead to various cancers. More importantly, we can screen drugs to see if they can prevent or cure the disease. This is a great segue for what I was going to tell you next. So we took these mice that have breast cancer and injected rhabdosin into the breast fat pad. We did that weekly and guess what? Our experiments worked." Sam looked as delighted as a child who had found his favorite toy. "Rhabdosin works, rhabdosin works, it cured breast cancer," he concluded, flushed and breathless.

Erin had eyed him with speechless wonder.

"Erin, you can't imagine how big a story this is. We are writing it up for publications in a high profile journal. Once this work is done, we are going to systematically analyze other human genes that have been linked to cancers such as BRCA1, BRCA2. This would be such an exciting time for the lab."

"What's next?" Erin had asked, still unsure if she should dare to dream along with Sam.

"Dinner!"

"No, silly! What's the next chapter in your research?"

"My dream is to see if rhabdosin will work in humans: Breast cancer, ovarian cancer, colon cancer, prostate cancer? Will it work for women with breast cancer for whom tamoxifen does not work? Will it work for women who have cancer-inducing mutations in BRCA1 and BRCA2 genes? These gene mutations make you susceptible to cancer. Will it work for other human diseases? Will it be the panacea for ending human cancer? Who knows? You never know until you try."

Sam had looked very animated that day. Clearly, he had been very excited about his work. Erin, although she tried not to show it, was thoroughly shaken up. On the one hand, she had wanted to ask him a thousand questions about cancer. At the same time, she did not want to share her inner conflict with him. She did not want him to know how she felt about him, how much she was starting to admire him. She went home and laid awake all night. When am I going to see him again?

Sam suddenly stopped coming to the literary sessions. Erin wouldn't admit it, but she had felt hurt and upset. She later realized that they'd never really made a date of meeting after the literary sessions. It had just happened that they continued their talks at the café, nothing preplanned, just a natural expression of a budding friendship. Or may be not friendship at all, just mutual interest in each other's profession. Erin toyed with the idea of calling or texting him, then stopped. Was this interest more from her side? Why couldn't he get in touch with her? Anyways, she ended up finding out from her friends that Dr. Sam Roy was very busy with a stupendous breakthrough.

She had always been an avid reader and sporadic freelance writer since her non-profit grant writing days, but Erin had been too lazy to find an agent to publish her creative endeavors. As her interest in attending the Thursday literary sessions without meeting Sam waned, she joined an online poetry blog, "As We Speak." Nature had always helped her cope and after a long weekend walk on a comparatively remote area of lake nearby she had posted a poem there:

I suddenly feel it ...
The peace and the gold of the sunset
seeping through my skin and flowing in my veins
as the sky and water merge in soft lapping waves of liquid color
with water geese swimming in unison without quacking
as if wowed into silence by the palpable serenity
content only in leaving wakes of rippling life stories
constantly changing shape and form, the only real picture on
the waves of time ...
As I stand alone at the pier head trying to heal and understand
watching wild flowers smile and nod at me, clinging bravely on
the rocky shore;
with water on three sides, earth on one, sky above
and life everywhere ...

Erin always wrote under the pen name Bonita, in her mom's memory, for Nancy had expressed her hope that her daughter would be a great writer one day.

One of the responses to Erin's poem on the blog read "I loved your poem. I recently lost someone very dear to me, my mother. Your poem really helped me. Please keep writing." It was signed, Sam. Erin wanted to ask Dr. Sam Roy, if this Sam was really he, but wasn't sure if she should bother him, busy as he was with his important research.

Then late one night Erin had been checking her emails in front of the TV when she saw an interview on the Medical Watch program. A medical researcher from Aurum University, Dr. Sam Roy, was talking about how he had been conducting groundbreaking tests in administering rhabdosin as a future cancer cure. His experiments on mice had been successful, and he was ready to do clinical trials on humans. Her Sam had become so famous. She felt proud and happy, only to cover her flushed cheeks the next instant, at her unexpectedly effusive reaction to the news. That is why he lost touch with me, she thought, with a strange mixture of admiration and hurt. Several contact numbers were flashing on the screen where eligible breast and ovarian cancer patients could call. Erin sat very still, her eyes glued on the screen. If only she had heard this from Sam when her mom was alive, maybe she would have been here with her today. She hurriedly brushed a tear away and thought rapidly, she could always give her aunt Angela this contact number and ask her to call Dr. Sam Roy for further advice if she needed to. Erin was just reaching for her phone when a thought struck her. Dr. Sam Roy had to be the same Sam who had commented on her poem on the blog! But she couldn't be too sure, Sam was such a common name! But Erin couldn't help wishing that it was her Sam who had commented! For then Erin, the working girl with immense aspirations and Dr. Sam Roy, the research scientist in the making, could meet and interact on equal terms, at some level. But it was also very unlikely for such a high profile medical researcher to spend time looking at some poetry blog. She turned to look a little more closely, but the news flash had ended. She called her aunt Angela right away and told her about Dr. Sam Roy.

Over the next few months Erin had to go out of town several times to meet prospective library clients and benefactors and also to attend administrative training courses. On her return she called Angela's cell phone and was thrilled to hear Kathy's voice. "When did you come back Kath? How long will you stay?"

There had been a short silence, a sharp sob and then the line went dead. Erin just thought that the phone had lost charge and that Kathy would call back. But Erin had had a long and exhausting day and so she fell asleep right away.

The next day, Erin was on her way to her weekend job at Dr. Lee's when aunt Angela called. Her voice sounded thin and withdrawn. "Erin, I can't find the contact numbers of the research scholar at your university, who you had heard on TV the other night. Would you mind giving them to me again?"

"Are you ok?" Erin had asked apprehensively.

"It's Kathy," her aunt's reply had a tremor.

"You know Kathy had gone to work in the West. But she'd had a lot of pain in the abdomen lately and one morning it got so bad she had to be rushed from her work to the emergency. They found out she has ovarian cancer, advanced stage. Kathy wanted to come back home to her family. So she's in the hospital here for further tests. The doctor there also gave her Dr. Sam Roy's name and said he is trying for an experimental rhabdosin cure that is already being hailed as a future cancer conquering drug. Remember you had told me about him being interviewed on TV? You know him, right? How do you think I should contact him?"

"Well I really don't know him … in fact haven't seen him for quite a few months now … but I'll find out all the details as soon as I get to my office. Have you talked with Dr. Lee, your general physician, yet?" asked Erin, stunned at this unexpected turn of events.

After reassuring aunt Angela, Erin was reeling from the shock of hearing about Kathy's cancer when she arrived at Dr. Lee's office for her weekend shift. She was told that Dr. Lee was doing hospital rounds, so she spent most of the morning reading up on Dr. Sam Roy online. Dr. Lee came into her office when he returned. "Did you see Kathy?" she blurted out. From the look on his face she knew things were not ok. Erin couldn't help complaining about her ill luck that all her relatives had contracted this "scourge of mankind."

Dr. Zen Lee looked pensive. Finally he said, "Erin, I don't know if you can call it ill luck, but sometimes family members inherit mutations that make them cancer prone. As your family doctor and a friend I would strongly recommend that you stop worrying about Kathy. She's here now and will get the best of care and I promise with her permission I'll get in touch with Dr. Sam Roy. But, you are in the same family, my dear, and I strongly recommend as your general physician also, that you get yourself checked right away. Let me do a blood culture on you. I'd like to send it to Myriad Genetics in Salt Lake City that has a number of patents for testing the mutations for breast and ovarian cancers. With the new Affordable Health Care act, there won't be any issues with insurance companies paying for the testing. I would like you to take care of it right away."

Erin was speechless. All along it was her family she had been worried about. It had not occurred to her that she was very much a part of this saga as well. Though hesitant at first, she eventually realized the urgency in Dr. Lee's suggestion and got herself tested.

Erin was on her way to work on the morning the Myriad test results were due, and her thoughts were broken by her phone signaling that she had a text. It was from Miguel, her junior colleague, telling her that Erin's two morning meetings had been taken care of, but a research scholar had been waiting for the last fifteen minutes and constantly enquiring how late she would be. He insisted he wouldn't leave without seeing her. Disgusted, Erin called Miguel and barked into the phone, "Miguel, please remind me to send a memo to the administration not to let random research people come in without appointments … Just who the hell does this person think he is?" Miguel interrupted her. "Erin, he said his name is Dr. Sam Roy and he really wants to talk to you. He thinks you may know someone named Kathy?"

Erin dropped the phone, then quickly picked it up and almost ran the rest of the way. Surely Kathy was ok! And … how was Sam? She hadn't seen him for ages! She skipped up the library steps and stopped for a few minutes to compose herself, before walking into her library office somewhat unsteadily through the side door, so that she could avoid the lobby where Dr. Sam Roy must surely be waiting. She quickly poured herself a cup of coffee in her office, took a few scalding gulps, rather annoyed at herself for being so keen on meeting Sam again after last fall and told herself repeatedly that the main reason was Kathy and her health, nothing more or … less. She then switched on the intercom. "Please ask Dr. Roy to come in." She decided to look and behave every inch an important official of a university library.

Erin remembered how professional Dr. Sam Roy had appeared in the television interview, very different from the Sam who had munched her silly little snacks after the literary sessions and declared them yummy. She felt rather embarrassed about her novice concoctions, but at the same time she was truly thankful that he wanted to talk with her about Kathy today. Had he seen Kathy already? Kathy must have told him she was her cousin and she worked in the library on the same university campus as his research lab. Dispersing her chain of thoughts, the door opened and Sam asked crisply, "Erin Goldberg? Remember me? I am Sam."

For some unknown reason Erin found herself suppressing a giggle thinking of Dr. Seuss's book *Green Eggs and Ham* where he says "Sam I am." It had been one of her favorite reads when she was a child. Nodding, she stood up, walked round her desk quickly and shook hands with Sam. If he noticed her formal greeting he expressed no surprise. His grip was warm and strong. He was shorter than Erin remembered, and he seemed thinner. Unlike before, he had round thin dark rimmed glasses and a deep golden tan. His shock of dark hair had grown longer and looked alive. He wore weathered jeans and a hooded sweatshirt with the university logo on it, not the twill pants, faded dress shirts and V-neck pullovers he had before. His faint Indian accent had acquired an interesting western flavor, like his whole person and look. Then Erin met Sam's dark eyes, as intense as before and chuckled inwardly. Yes, Dr. Sam Roy had turned into a slightly older exotic version of Harry Potter.

"Just let me double check. You are the Erin Goldberg, Kathy's cousin, right?' he asked decisively, emphasizing her name.

"Was my aunt able to contact you in time? Have you met Dr.

Lee yet? How soon will your new medications make Kathy get better? What is the name of the medication you will give my cousin, Dr. Roy?" She ended in a gush and then chided herself as she saw Dr. Roy's face lighten. Was he thinking she was behaving like a little kid? She'd have to maintain her professional calm in future. But how could she? Wasn't this a life and death issue for her cousin? The questions flew out of her quite endlessly.

"Oh come on Erin. You have always called me Sam. I'm sorry, I should have texted you or something. Actually, I suddenly got too busy to attend the literary sessions, but I do miss them. How are they going?" He seemed as easygoing as ever. So much so that she almost told him that she too had stopped attending the Thursday meetings after he stopped showing up. That would be a little too forward, wouldn't it? For all she knew it may make him disappear again! She smiled noncommittally and said, "They are going ok, Sam."

Missing her hesitation, Sam continued, "Erin, please don't worry about your cousin, Kathy. Actually I didn't come here to talk about her only. She will enroll in my clinical trial if or when I am ready with all the approvals. I have come to talk to you about yourself and what your family history means for your future. After checking your cousin's health history, knowing that your mom passed away from breast cancer, I feel that there is a strong likelihood of a family history for mutations in genes causing breast and ovarian cancer. I know there are some special tests that can tell you if you have such inherited mutations. In fact, I insist, you get them done ASAP, for old times' sake."

Erin was outraged. Barging into her office after all this time, crashing her late morning schedule, all of this she had tolerated for Kathy's sake. She wanted her cousin to get every possible option to be cancer free. But Erin had no intention of seeking Sam Roy's uncalled-for advice for herself. Whatever her health condition she would like to make her own decisions on her own accord and not be pushed into something by an over-confident stranger.

Dr. Roy was almost that, wasn't he? He was not her physician, not her adviser, just a researcher working in a high profile field, in the same university as she. Composing herself, Erin replied, "Dr. Roy, I mean Sam, I have heard about your great cancer research, but right now I would like to concentrate on Kathy only."

Sam looked at Erin, smiled briefly and replied smoothly, "I'm sorry if I upset you. I do apologize I didn't get in touch with you earlier. But you know me Erin, I am rather impatient and can't waste time to come to the point. I'm not very good with niceties either. Do you think I could sit down with you at a convenient time to discuss your health history as well?"

Though she refused to admit it, Erin was impressed yet again by the way Sam's smile livened his expressive face and surely by his graciousness and insight. She also remembered how she had taken an instant liking to the way he pronounced her name, kind of rolling his tongue on the *r* and stretching the *i*; "Er-een." His enigmatic eyes were quizzing her face.

As she was about to answer him, there was a brief knock on her office door and Miguel stuck his head in. "Sorry Erin, Dr. Lee is on the line and wants to talk to you right now. He says it is urgent." Erin turned to Sam to end their meeting but he was glancing at his cell phone, and before she had time to word a reasonably polite goodbye, she heard him say unhurriedly without looking up, "I just texted my office I'll be out for lunch. So I think I'll wait here till you come back. I hope you won't mind?"

"As you wish," she countered quite irritated, her mind already on what Dr. Lee was waiting to say. Erin took the phone call in the empty conference room.

Dr. Lee said, "I don't know how to put it Erin, my dear. Normally, I would not give you this news on the phone, but because I have known you since you were a child, and you know I have your best interest at heart, I will give you the lab results. Let me just tell you, you are young and strong and since you know beforehand you can fight and win."

Erin cut in, "Dr. Lee what are you talking about?"

"Erin you have tested positive both for BRCA1 and BRCA2 mutations," Dr. Lee answered shortly. Erin didn't know how to take that. Should she say "thank you?" Perhaps. Because hadn't the doctor's advice led her to this early gene diagnosis? But then why did her legs suddenly feel so weak and the bright sun outside seem so dull and listless? All this time she had been a bystander, an observer to the whole family drama but instantly she was very much a key player. She felt a mountainous wave of self-pity engulf her and she had no clue how to keep her head up, whether to laugh or to cry.

Dr. Lee was saying the report said she had a 80% chance of developing breast cancer. But there was still that 20% chance of never getting it, right? What if she were lucky enough? Erin thought, grasping at straws, what if? Or maybe she could go in immediately for double mastectomy like her aunt Angela through the True Pink organization to which she donated ever since her mom passed away. Whatever it was, she knew she would have to fight it every inch of the way. Actually, she had no other option but to fight this inherited mutation. She had not been able to escape from her toxic past, but she always believed that she would emerge from it a much stronger and more real human being. She had vowed at her mother's bedside never to succumb, to try her level best to conquer whatever odds life threw her way. She knew it would not be easy. She had to find ways to be aware of her choices and make the right decision at the right time. She only had to know how and where to start. She had to do it fast, before it was too late. She could not wait any longer.

With this sudden rush of adrenaline and an acute sense of urgency rather than dread, Erin returned to her office. She stood for a minute dazed to see Sam still sitting there. For her a lifetime seemed to have passed since she had gone to take that call but it was barely minutes.

Sam observed, "That was quick." Then he looked up and on

seeing her face walked quickly over and lightly touched her elbow. "Are you all right, Erin?"

"When can I see Kathy?" answered Erin trying to take her mind off herself.

"Let me repeat, Erin" Sam said patiently, "Kathy is now in capable hands. It is you I want to focus on. I can't emphasize enough the threat of your inheriting problem gene mutations, Erin, and what your options are, to be cancer-free," Sam ended quietly. The impatience and brusqueness seemed to have left his manner and voice. It is easy to fight rudeness but not concern and Sam's seemed genuine. Erin couldn't take it anymore. She had to turn away to the window. "I'll think about it … later and let you know," she said coldly and tried to dismiss him by turning her back on him.

She stood for a while looking out on the college football field. There were students running along texting or talking on their phones, teachers carrying computer bags walking purposefully, friends chatting in groups, everything looked so normal. It was only she who was abnormal and who didn't fit in this scene. She suddenly seemed an alien in her very own work place, her comfort zone, ever since her mom had passed away. Everyone would pity her if they came to know. Covering her mouth to suppress an unexpected sob, she quickly spun around from the window to walk away but found herself enveloped in a warm hug. Sam had not left yet.

"No, no!" her head screamed as she tried to extricate herself from Sam's steadying embrace but her body and soul craved for a little sympathy. If only her mom were here now, or even a close friend she could confide in. She couldn't burden her aunt Angela anymore. Erin's body gave a little shudder.

Sam patted her shoulder gently and murmured, "I won't ask what happened or what you heard just now … but you know I'm here if you need to talk. For now, just let it out," he added. And then Erin, the smart, practical, efficient administrator buried her head in Sam's unexpectedly stable shoulder and sobbed inconsolably.

After a couple of minutes she moved away rather self consciously, looking for a tissue on her desk. Sam brought the tissue box over as he helped her to a couch nearby. To diffuse her embarrassment and the tension in the air, Sam stole the confusion out of her, by saying briskly "Everyone has their moments, so no worries … it's quite normal."

Erin's self pity spilled over. She said shakily, "But that's just it, I'm not normal. I'm …"

Sam cut her short, "Whatever it is, please let me help, I'm your poetry pal remember, did you read my response on the poetry blog complimenting your poem, my dear, Bonita."

As Erin looked up in pleasurable surprise, dabbing her teary eyes, that it was indeed this Sam who had complimented her writing, he gestured at her necklace. "I saw it just now, so you are my favorite poet 'Bonita' from, 'As we Speak'? Didn't I write to you I had lost someone special too? Yes, like you I lost my mother to breast cancer, and that is why I've come so far away from the country where I was born. I will not quit till I have found some medicine, some means to conquer cancer. To me, it's a reality in the making." The conviction in his voice was not conceit, it sounded like a promise of powerful hope and Erin found herself looking into the kindest, darkest, most resolute eyes she'd ever known.

Then she suddenly laughed out loud and said, "Sam, I thought you had forgotten me" and felt herself blush immediately at her own awkward telltale comment.

But Sam rather unexpectedly and strangely to her liking, answered, straightening his sweat shirt and looking aslant at her, "Really? Then we just have to meet more often to make up for that. No time like the present, Bonita, to start. So please let me take you out to lunch. Then if you wish you can tell me what's on your mind."

"Ok," said Erin instantly and wondering if her acceptance had been too fast, added, "But not today, Sam, for I have a very full afternoon."

Erin had to admit she felt much better after that good cry. "I will talk to him later," she thought. "After all, what better friend than a medical researcher to help her battle her health issues." But she was not sure if she wanted to tell him her whole life story so soon, so fast.

They decided to postpone their lunch date and schedule it later when Erin felt up to it.

two

Nick's story: Colors of hope

The smell of chocolate drifted through the whole house. For the last nine years, Paula had been baking this same cake on this day. Sitting in front of the gleaming kitchen countertop she looked up. She needed to take the cake out of the oven in two minutes. It was late afternoon and the kitchen was flooded with sunlight. Rebecca's photo looked almost unearthly in the glow of the afternoon sun. Sometimes, Paula wondered what things would be like if events were different.

Rebecca was Paula and Jerry's only daughter. She was tall, academically bright, a good athlete, and an excellent violin player. Images after images flashed through Paula's mind. Five year old Rebecca, Becca they called her, coming home with her first place trophy that she won at the swim meet, ten year old Becca making everyone spellbound by playing violin at her friend Kathy's birthday party, Becca leading the school basketball team at age 14 in the interstate basketball competition. Their only daughter! Paula remembered her first word, her excited face the day she stood on her legs without holding onto anything for the first time, the first book she read to her. Memories brought tears to her eyes. How was it even possible that Becca, bright, strong, beautiful Becca, had been diagnosed with the fatal disease of Hodgkin's Lymphoma.

Becca had run away with her boyfriend in her senior year of

high school, shocking all those around her. She came back home one December evening looking pale and tired. Within a few months she gave birth to Nick. Nobody came to know who Nick's father was. When Nick was six months old, Becca passed away at the age of 19. Her life was wasted before it had a chance to bloom. Paula and her husband Jerry took the responsibility of raising Nick. Paula kept on dusting Becca's trophies, books and violin. She did it every day. Somehow, it gave her a feeling that her daughter was still there.

Paula closed her book, took off her reading glasses and walked to the oven to turn it off. Her train of thought was broken at the sound of her husband's footsteps. He had just come downstairs.

"I'm worried about him," Jerry told his wife.

"What's he doing, he should be coming downstairs by now. His friends are supposed to be here in half an hour," Paula said. "I don't know, I tried to check but his door was closed."

Paula frowned. She knew Nick was fighting another bout of fever and chest congestion. He was upset that he missed the basketball game last Monday because of his health. He also was looking more pale than usual. It was taking him longer this time to get better. This reminded her of his childhood days.

As a baby, Nick used to get frequent infections of all kinds. He kept Paula and Jerry so busy that they did not have time to grieve over the death of their daughter. Paula played the mental slideshow again: two year old Nick fussing with high fever, three and a half year old Nick coming home early from school with his legs swollen, his nose bleeding profusely, eight year old Nick throwing a huge tantrum as he missed the swim meet because he was diagnosed with strep throat the day before. In the beginning they thought that Nick would outgrow all these health problems. Instead, his health only got worse. As a 12 year old, Nick missed more than half his school days. He could not participate in a lot of physical activities. They had decided to move to the West Coast in the seaside town of Bayport, thinking the sun and ocean breeze would improve Nick's health, but there was no such luck.

Paula went upstairs. Nick was not in his room. It was almost time for sunset. Soon the western sky would light up with different shades of red, purple, and pink. Nick sometimes went to the viewpoint across the house to see the gorgeous sunset over the Pacific. Paula noticed the intent gaze on Nick's face when he looked at the Pacific down below. The turbulent water at times reflected the chaos of his emotions but, like the sea, they were at times unpredictable.

Later, Nick went to the viewpoint to watch the sunset, and he went alone. His friends were either riding bikes or playing at the beach. Nick could not do either. He could not enjoy the outdoor activities that his friends did. He felt the ocean breeze on his face. It made him restless. He was often disturbed by thoughts that not only his teenage years, but also the rest of his life would be spent the same way. Nick knew that his grandparents were extremely concerned for him. He felt bad for them. But he was trapped in his own helplessness. Looking at the ocean and the fury of its churning waters he wished he could have an iota of that tremendous life force to fight off all the diseases. He had been diagnosed with rhabdomyosarcoma a few weeks ago. A small mass on his left leg had been detected. A biopsy had confirmed the initial suspicions of the doctor.

Right before he left the house he had heard his grandparents engaging in a deep conversation. It was about a blood test for him that one of his mother's school friends, Kathy, had suggested. Nick recognized the name but barely remembered Kathy. He did not have any memory of his mother so he did not pay much attention to Kathy when she came to visit them in California five years ago. However, now that he was older he was curious about her. What was his mother like? He knew that his grandparents felt strongly about Kathy since she was their only tie with their daughter, besides Nick. She was the long-lost school friend of their daughter. That was the only reason they invited her over today, as soon as they came to know that Kathy was in California at that time. To-day was a very special day and Nick's grandparents were planning a very special party. Today was Nick's birthday.

The sun went down. The sky still looked glorious, and the ocean water reflected the colors. But Nick knew from his experience that the colors were all transient. Within a few minutes the colors would disappear from the western horizon and both the ocean and the sky would start looking black. Nick walked towards his home. Even that short walk made him exhausted. As soon as he entered the house his mood was lifted. There were balloons everywhere. All the lights were burning brightly. He inhaled sharply and took in the aroma of his grandma's baking. His friends were not there yet. But as he entered the kitchen he met a lady who was sipping some kind of herbal tea, sitting at their kitchen countertop. He figured that she must be Kathy.

Kathy was tall and skinny, a little too skinny for her build. She seemed to be in her twenties. But her face did not reflect the glow of youthful liveliness. Her hazel eyes were bright and shiny though. Nick knew that Kathy was fighting the terrible disease of cancer. Still, she still managed to look bright and cheerful. As Nick greeted her, she immediately rose to her feet and came to give Nick a quick hug. Instantly Nick stepped back. He was already struggling to breathe. He did not want to come too close to anyone. He just gave her a quick hug and went to answer the door as the doorbell rang. On his way to the door he caught three or four phrases uttered by Kathy: *Chicago, oncologist, great doctor, Sam Roy.*

Nick opened the door. It was one of his classmates. Eventually all of them arrived, all eight of them. A group of teenage boys, full of energy! Nick took them all to the basement to play games. Nick forgot his physical limitation and joined his friends. It took him only a few minutes at the table tennis board to collapse on the table wheezing. He could not catch his breath. His friends ran upstairs to get his grandparents. They rushed to the basement with anxiety etched all over their faces. Kathy followed them too. They all helped Nick upstairs to his bedroom. Grandma gave him water and a pill. He did not care about his medication anymore. All he knew that it would help him fall asleep soon. By the time he woke he would feel better. But for how long? Was there any long lasting solution to his ongoing health issues?

Before passing out in his drug-induced sleep he heard Kathy suggesting his grandparents come to Chicago to see Dr. Sam Roy, the famous doctor, who has been treating Kathy for her ovarian cancer. Who was this Sam Roy? Nick drifted off to sleep. He dreamed about Sam Roy that night. Another sterile room with cold walls? More needles? Another serious face with grave eyes telling him what he could not do because of his health condition? The colors of sunset started fading out.

three
Sam's story: Man with a mission

Sam was a brilliant student all his life and hailed from a middle class background. He studied in a well-known school in the southern part of Kolkata, a bustling city of 15 million people in Eastern India, in the state of West Bengal. Kolkata was formerly known as Calcutta, the name given by the British rulers who had ruled India for more than 200 years.

He went on to obtain his medical degree in Kolkata, with a specialty in infectious diseases. He received a scholarship thereafter to do his PhD degree from a well-known US University in Chicago. At a very young age of 24, he had done a phenomenal job in contributing to the revolutionary discovery of genome sequencing. He was summoned by several world class universities to give talks in his area of expertise. After completing his education, he joined a famous medical research institute and hospital as an infectious disease specialist with interest in oncology. This institute was affiliated with a university in Chicago where he was a clinical research professor. His professional activities included not only to treating patients, but also actively engaging in translational and clinical research on cancer prevention and drug development.

Sam was originally named Sambuddha Roy, but his first name was shortened to obviate futile attempts to get the correct pronunciation from struggling colleagues. He was a renowned scientist by his own rights, and he had coauthored many books and research publications with his students and colleagues.

There were two different triggers that strengthened his resolve to become a world-class scientist.

The first trigger was in 1990. Sam was known as Amar then. All Bengali children have nicknames. *Amar* means immortal or one without death. He was twelve years of age and a student in 7th grade. His father, Prasanta Kumar Roy, known as Mr. P.K. Roy, was an assistant technician in a testing laboratory in Kolkata. The owner of the laboratory was a multimillionaire who was also a renowned pathologist. The name of the laboratory was Roy and Ganguli Diagnostic Laboratories. This was a renowned pathological laboratory in that area. For quite a few years of his childhood, Sam had the misconception that the *Roy* in the banner of the diagnostic center stood for his father's name. Much later he had found out that the Roy was actually the person who had collaborated with Dr. Ganguli initially and who also co-owned the diagnostic center. Sam's father was an absolute hero to him, and remains a hero to this day. He was also a hero the day he saw his father sterilizing the test tubes and distributing them across the different tables or when he saw his dad labeling the blood filled test tubes and carefully storing them in some kind of machine. He had seen his dad take orders from junior doctors or even interns. That day was a bit of a revelation to Sam, but since his dad had never lied to him about his job, it only glorified him in Sam's eyes. Sam used to bug him as a child on the details of his job and his dad used to smile and answer. "I track those bloody bacteria and viruses and help doctors to find a way to kill them," which was technically true to every letter. Sometimes, he used to study the bacteria under the microscope. His father had been there in the lab for fifteen years and had acquired a decent level of workable knowledge on many bacteria. He could tell a *Pseudomonas* from a *Bacillus*, or a pox virus from an influenza virus. Mr. Roy had the habit of going to College Street and buying used books on bacteria, viruses and pathology at very affordable prices. His dad had wanted to be a scientist, but due to extreme penury and lots of financial commitments towards his sisters and ailing parents, his dream was never realized.

Sam and the lab owner's daughter, Sonali, often met each other when Sam went to meet his father. Sam's home was hardly three blocks away from the lab. That is why he often went to see his dad after school—to roam around the laboratory or to be seated in Sonali's dad's chair (in the command room). They often played outside in the lobby when they were kids. She was a year younger than he, and was a grade lower in a different school. Sam was instructed by his dad to behave himself especially well when he came to the lab. His dad's interest in explaining strains of viruses to him or to give him a peek at some bacteria used to diminish when Sonali was around. It was as though this little knowledge transfer to Sam was not really a cool thing, in the environment of that diagnostic center. It was on one of those days that the disaster happened.

Sam had always watched Dr. Ganguli's Mercedes S-class car with awe. He never mustered enough courage to touch it. Every now and then when he was drawn close to the car to take a better look, he had been yelled at. It was as though there was an invisible electric sensor or a magnetic field all around this car, so that whenever he was less than five feet away from this car, someone yelled at him to stay away. Either it was the driver, who was merrily smoking a cigarette outside the car, or the gardener of the property or the gatekeeper. One afternoon, the driver had carelessly gone somewhere nearby, maybe to attend nature's call. The door of the car was not locked and the window pane was pulled down. Sam immediately darted towards the car and stopped as he approached five feet from the car. There was no yelling noise. He tiptoed to the door, opened it and then sat on the driver's seat. He had no clue as to how to drive any car, let alone a posh Benz. He started the ignition and the engine revved softly, much more softly than the bulky Ambassador model cars that punctuated every street of Kolkata with their characteristic yellow and black colors. He did not know how to handle the gears, but started playing with the gear shift, and lo and behold, after a few seconds the car started to move. He was shocked. An adventure to "feel" a driving experience had suddenly spiraled into a sinister irreversible process. He shouted for help as the car approached a banyan tree in the garden. The gardener shouted in dismay. The car was now going at about 15 miles an hour and before he knew anything, the bonnet of the car hit the banyan tree!

Smoke oozed out from the crevices beneath the bonnet. Sam was petrified. His forehead hit the steering wheel with a thud and he shouted for help. The damage was already done. Before he could comprehend anything, the driver, who was now back, pulled him by the ear and took him in front of Dr. Ganguli. There were drops of blood that he could see trickling down his cheeks and drenching his white school shirt. His father came running and started to wipe the blood from his forehead. The afternoon was hot and sultry and the smell of blood mixed with his own sweat is something that got imprinted in Sam's mind forever. Dr. Ganguli was briefed of the incident by the enterprising gardener, some onlookers and overenthusiastic witnesses who immediately conjured different versions of the story to impress Dr. Ganguli. Dr. Ganguli just uttered two words almost like a whisper which he did not understand. "Namak Haraam" (which summed up to something like disloyal and ungrateful). His father came pleading with folded hands and said "Sir, he is just a kid, please forgive him. I will admonish him."

"What the hell will you explain to him, that the bonnet will cost three times your monthly salary to repair?" Dr. Ganguli was fuming. He walked towards his father, who was very apologetic and standing with his head hanging down. Dr. Ganguli went on with his tirade. "Roy, you have been our old employee and I always liked your work, but this kind of nonsense is not tolerable. Please teach your son to stay out of our building." Sam's dad had always taught him to respect Dr. Ganguli, but this was a completely different picture of Dr. Ganguli that emerged that day.

One of the colleagues even suggested "This kind of anti-social kid should spend a few nights in jail to be taught a lesson." It pained him to see his father getting insulted like this. His father's face was very distraught and had a mixture of fear, apology, and helplessness.

"Sir, please deduct the amount from my salary in the course of the next six months, but please don't hand him to police. It was a stupid prank. I will explain to him myself. Please don't punish him, sir." His father was begging Dr. Ganguli with folded hands. This was not the same smiling dad Sam knew, a confident hero who was always in command.

Dr. Ganguli gestured angrily with his left hand as though he was shooing away some crows, "Please disperse … all of you."

Sam suddenly realized the enormity of his blunder. This was perhaps the worst incident that he had encountered in his life of 12 years. They walked out of the office and hired a rickshaw for the short distance home. His dad held a handkerchief tightly to the bruised portion of Sam's forehead and went on crying. Sam did not know how to react, but he found himself helplessly crying too. His dad rebuked him affectionately "Son, you should never have touched that car. They are very rich. They are different." Sam knew the meaning of different. His father carefully placed the word rich and different in two sentences.

Still crying, Sam told himself, "I will have my own Mercedes and I will have a much bigger diagnostic center than Dr. Ganguli. And I won't be different." It got imprinted in his DNA right at that moment, if there is such a place as a promise made to self in DNA structure. For the first time, he understood that his father could be so powerless, and could take so much insult just so he could have a job.

His dad resigned, much to the shock of Dr. Ganguli. They stripped him of his bonus and many other arrear payments. But the insult of that day was something that the whole family could not digest. Many years later, Sonali had come a few times to ask for his forgiveness. But Sam had an axe to grind and being the kind of person that he was, he never forgot insults. He said to her face at the age of nineteen "Ask your big dad to come and apologize to my dad and mom." To this, Sonali had shed tears and gone away. Her few letters were either hidden or trashed by Sam's mom or Sam himself. In spite of the fact that Sonali was beautiful, and Sam liked her a lot, they did not see each other again.

four

Sam's early years and venture into a research career

O Sam was 20 years of age, studying in Kolkata as a medical student, when the second trigger happened, which increased his determination to become a medical researcher. His mother had not been doing too well. His father had left Dr. Ganguli for a slightly higher paying job, far from their home. His mother had been having frequent fevers and some blackouts lately. She was 43 years of age, and after a few rounds to physicians one of them suggested doing a mammogram and some specialized blood tests.

The tests were done as a routine procedure and life went on as usual. That evening, there was a call from the lab where all the tests were done. Sam attended to the call. It was a baritone voice that told from the other side of the phone, "I am Dr. Das from Apex Diagnostics, Park Street. I would like Mr. and Mrs. Roy, your parents, to visit me again and have a consultation."

"What is it, doctor? I am their son. Dad and mom have gone to watch a movie. They will be late. I can take a message," Sam went on.

There was an ominous pause from the other side.

"Sorry, I am not allowed to speak to anyone other than Mr. Roy. I would request your father, Mr. Roy to visit me alone between 3 and 4 pm tomorrow, it is urgent."

Sam had heard about the visit of his mother for some tests but was not sure of all the details. The doctor would talk to his dad alone in confidence.

At 10 pm, when Sam's parents returned, Sam looked at his mom with a different eye. She looked so happy and pretty! At 43, she looked ten years younger. In fact, if dressed properly, she could pass off as his elder sister.

His mother exclaimed "You know Amar, I saw this beautiful daughter of Mrs. Sen at the theater. Wow, she would be a perfect match for you and you know, she is also great in cooking. Seema told me that. Why don't I call them for dinner at our place …" she went on and on. Somehow, Sam's mother was an archetypal Bengali mother who just dreamed of Sam finishing his medical degree, getting a job, and then getting married. She was more concerned about Sam's health and well-being than his grades in class. Fortunately, grades were the least of his parents' worries. Sam was always at the top of his class. His dad had also been an immense help in his math and science lessons whenever he got stuck. Sam had almost forgotten about the call, until the phone rang again from someone else and his dad answered the call. He told his dad immediately after the call "Baba (Dad), Dr. Das from Apex diagnostics called you. He wants to meet you in person tomorrow between 3 pm and 4 pm. Is it something serious?" His father suddenly became solemn. He sat on the sofa looking towards the floor. He asked Sam again. "Are you sure it was the doctor, or was it his assistant?

"He said Dr. Das … kind of baritone voice."

"Do me a favor, will you? Please don't share anything about this phone call to your mom and Ria." Ria was Sam's younger sister.

"Is it serious, Baba?" Sam sat beside his dad.

"I don't know," Mr. Roy said and started watching TV, sifting through channels like an inattentive child.

The next night Sam came to learn that Dr. Das had interpreted the mammograms and confirmed the presence of a tumor in his mother's breast, and that would they have to perform a biopsy as soon as possible, just to make sure. What followed during the next few months was traumatic to the whole family. Mr. Roy took a full week off after the biopsy results to communicate to Sam's mom that she had a malignant tumor and that it was in stage III. This drove the family to despair. Sam tried desperately to find second opinions from two other renowned doctors. The professors of his university loved him for his academic brilliance and diligence. They had referred him to their childhood friends who were oncologists in different medical schools. He also consulted with several renowned oncologists across the globe.

Sam and his sister Ria had struggled to keep smiling in front of their mom. In the last stages after her chemotherapy, when she had a bald pate and patches of blue color in her skin, she became weak and increasingly irritable and lost her temper many times on small things. When the ceiling fan was not set at maximum speed, or if the TV volume was too high, or if the food was a tad hotter than normal, she started complaining. She had become too frail to tolerate anything not normal. Was she not geared up for this journey? Even after the mastectomy, radiation, chemotherapy and doses of tamoxifen, there could be nothing that could bring her back on track. Sam got restless. He started corresponding with oncologists abroad in the UK, US, Germany and Switzerland, and also reached out to a few senior medical students who were in those countries for their higher studies in medicine. Could there be anything that could attack only those miscreant demonic cells and leave the normal cells at peace? Doctors had detected that metastasis had already set in his mom's bones and the bone marrow test had also confirmed the same. Could there be any cure? All he got was lukewarm responses.

He even went to the head of his medical department in a fit of rage and told him "What is the use of all this research if we cannot attack the malignant tumors separately?"

He had broken down in front of his oncology professor. His professor said "Sambuddha, you have a long way to go. There is nothing much we can do for your mother. This is a government sponsored medical school with a tremendous uphill task to even get an old microscope replaced, let alone to buy a new PET or a CT scanner. You must go to the very best places of research in the world and keep working. Many millions of mothers will be restored to their sons and daughters."

At that moment, he only cared about his mother. The mother that sang to him from the kitchen while frying fish for him. The mother that was a gentle shadow after a grueling day's work. The mother who diligently woke him up at 4 am on the days before his exams and also kept her TV volume mute while watching her favorite soaps, just so that he could study undisturbed in their small home with two bed rooms and a single bath.

The day his mom passed away with a smiling face and with two children and her husband on her bedside, Sam made another vow to himself "I will see the end of this road. I will go to the very best of institutes to do research on cancer."

Once Sam completed his medical degree in Kolkata, he did an internship in infectious diseases in Kolkata. He wanted to do a research fellowship in a medical school in the United States to learn more about cancers, particularly breast cancer. His mother's suffering had a lasting impression on his mind. Even though he used his familiarity with bacteria and viruses, that he learned as a kid in his father's workplace, and when he treated his patients with infectious disease, his interest moved more and more towards using his infectious disease knowledge to treat cancer.

This concept came to his mind when he read about the pioneering work of Dr. William Coley, a head and neck cancer surgeon at Memorial Hospital in New York, on the ability of infectious bacteria to allow cancer regression in his patients. This work was published by Dr. Coley in 1892, and Sam read with great interest the book 'A Commotion in the Blood' authored by S. S. Hall which described many of his successful cancer treatments with live bacteria. When 312 patients with sometimes inoperable cancer were deliberately injected with some of these live bacteria, 190 patients showed significant regression of their tumors. Occasionally, though, the bacteria would induce strong immune reactions, making the patients miserable or, in rare cases, leading to the death of very sick patients. Sam was well aware of the cancer-regressing activity of some bacteria such as *Clostridia, Listeria, Salmonella*, etc. He knew that physicians ascribe such bacterial cancer regressing effects on the activation of the immune system that occurs when a person gets infected with a foreign agent such as a bacterium. The concept that bacteria could actively fight cancers by producing anticancer agents such as a protein or a small molecule, similar to an antibiotic, was absent.

As an infectious disease specialist, however, Sam became deeply interested in cancer during his post-graduate studies. He wondered if infectious bacteria, some of which can live in the human body for months after infection, would actually fight cancer by producing such anticancer agents.

Because his mother's breast cancer was metastatic and spread to her bones, contributing to her early demise, Sam was particularly interested in understanding the ability of some infecting bacteria to fight metastatic cancers. It was during his postdoctoral studies in Chicago that he learned of some studies conducted in his neighboring university that bacteria such as *Pseudomonas aeruginosa* produce certain proteins that can enter cancer cells preferentially, but not normal cells. Such proteins attack cancer cells on multiple fronts. Even more interesting was this group's finding that these protein can also interfere in the invasion of the immune cells by the HIV/AIDS virus HIV-1, thus potentially acting as an anti-viral agent as well. Guided by such studies, Sam looked for anticancer proteins in many infectious bacteria, and as luck would have it, came up with one which he named rhabdosin.

Attending a cancer meeting in Chicago organized by the American Society of Clinical Oncologists (ASCO), Sam became aware of the fact that a company, with a license from the neighboring university, conducted Phase I human clinical trials in 15 stage IV cancer patients, with a peptide called p28 (because it contained 28 amino acids). According to the speaker, p28 showed very little immunogenicity or toxicity in these very sick patients, unlike chemotherapeutic agents. It also showed some significant beneficial effects such as stabilization of the tumors in some patients, partial regression of the tumors in two patients and, most importantly, in two patients complete regression of the metastatic solid tumors that were resistant to all conventional drugs. Sam was impressed at the ASCO talk. Through the literature, Sam also found out that besides the 28-amino acid long p28 fragment, there are bacteria that have been reported to fight cancer, and other bacterial proteins, not having anything to do with p28, such as arginine deiminase, have been shown to have anticancer activity.

Sam smiled at the ingenuity of these bacteria to design protein weapons to protect their homes—human bodies—against invaders such as cancers. At the ASCO meeting in Chicago, as Sam was talking to several attendees in the session, he learned that the physicians who presented the Phase I clinical trial results in 15 stage IV cancer patients chose to use p28 because it can be chemically synthesized, and therefore falls under the category of small molecule drugs. Because proteins are too large to be chemically synthesized, or would be very expensive to produce chemically, they have to be isolated from bacteria, perhaps with traces of toxic cellular contaminants. Thus proteins are labeled as biologics and will have to go through more stringent regulation by the Food and Drug Administration if used in clinical trials, making the trials costly and time-consuming.

Sam saw an opportunity. Why use only a part of an intelligently-designed bacterial protein weapon with multiple attack modes against cancer, when one can use the full-fledged weapon? Why not use a protein such as rhabdosin that showed both cancer therapeutic and preventive activities in his cell culture and animal studies, and go through a more stringent regulatory process, rather than take a short-cut and an easy route? Most important for Sam was not only p28's preferential entry into cancer cells with a strong attack when inside the cancer cells but its ability to prevent cancer in mouse mammary cells, as he read in the literature. He found out from a patent and literature survey that certain carcinogens can induce what's known as pre-cancerous lesions in mouse mammary cells. These pre-cancerous cells have the potential to become tumors. He learned from the literature that p28 can strongly inhibit such pre-cancerous lesion formation, thus potentially preventing cancer.

Of course, if some of these pre-cancerous cells escape p28 action to become full-fledged cancers, p28 can attack them by entering inside and killing them, just as he was trying to demonstrate for rhabdosin. The first thing Sam needed to do was to approach some potential investors to fund his research and clinical trials with rhabdosin, since he knew from experience that academic funding was highly competitive and hard to obtain for practical applications. Fortunately, he was in the United States where entrepreneurship was encouraged and where some investors were willing to invest in risky unpredictable projects.

Sam joined the staff of Aurum University, a small university in Milanburg, a suburb of Chicago. Sam's university was in close touch with Bill Harrison, the CEO of a large investment company in Chicago with an interest in the development of new generations of drugs. Since rhabdosin's effectiveness in allowing tumor regression in mice was already known, all he had to do was to find its toxicity in various animals when given in large doses. Of course he also had to determine rhabdosin's ability to prevent cancer induction in mice when given orally with a known carcinogen. He could get started as soon as he received a couple of hundred thousand dollars to buy the mice and various chemicals. Through the help from his university, he approached Bill Harrison, who was intrigued enough to invest a half-million dollars in the project.

five

Sam and his life at the university

It was mid May and there was pleasantness in the air. Sam got out of his bed on a Friday morning. There was this vision of a girl that he could not get his mind off. Her words were stuck in him. An American girl in her mid twenties who was giving a talk on medical ethics the other day in the district library. Her name was Erin. She was involved in the same literary group, "Beyond Words," which Sam had started to frequent every Thursday evening. What interested him was the way she addressed the institution of religious bodies and "moral police" as she called them and how they interfered with science in the name of ethics.

He was pondering her words. This lady apparently quoted profound Hindu philosophies as depicted in "The Gita" which happened to be a very heavy chapter in the epic, *Mahabharata*. She cited the example of Lord Krishna instructing the ace archer and fighter Arjuna to fight a deadly battle against his own cousins. What appeared as a blasphemy to Arjuna was just a small picture in Krishna's world of ethics. What Lord Krishna viewed was the ultimate good for mankind after the lesser evil had been vanquished. She also quoted the lovely words "What a caterpillar calls the end of the world, the master calls the evolution of a butterfly,"

True indeed. Erin's steady blue eyes were deep like the ocean, yet her voice had a resolve, a certain degree of thunder that comes from deep within.

Sam stepped out of his home and saw something that struck him. A man in his late fifties or early sixties looking with a fixed gaze at his house. This was not the first time he was seeing him. Stalker? Burglar? Nope, may be not. From a distance the person seemed to be of Indian origin. He wore a red cap and a blue windbreaker. He must have come from India to visit his son or daughter. This is the time when the neighborhood got flooded with parents of Indians flying in from India, to escape from the Indian summer and enjoy a few months here. What struck him as a bit of a coincidence was that this same person seemed to be looking at him every day for past few days at this same time, while he took out his S-Class Mercedes from the garage of his plush home in Hoffman Estates. On his way to the office, Sam always bought a cup of coffee from a local donut shop. The coffee was not a name brand, but he knew the owner who was an Indian and he had kind of gotten friendly with him over the years.

"Hi Sam, would you like our latest special chai latte?" The person at the window asked him amicably.

"Yeah, sure. Any chai would do." Sam smiled back.

"As a special promotion, we will give it to you for free the first time."

As soon as his car was back on the street, and Sam took his first sip, he was pleasantly surprised. This was indeed his favorite tea from his adolescence and adulthood, chai with cardamom, a touch of clove and even smelling of ginger. It tasted exactly like the tea that he was so used to having at the end of his gully (small road) in South Kolkata where he occasionally had a chit-chat with some friends. How could they make it here? It was normal for many donut shops to hire Indians as cooks. But was this person from Kolkata? He must find out tomorrow.

At his office the first thing that caught his attention was a yellow post-it slip on his desktop that said "See me soon, Urgent. BH."

The initials of Bill Harrison were familiar to him and he walked past all the aisles to reach the research sponsor's spacious office that boasted of many memorabilia that cast an aura about Bill inside the room. Bill visited the university every other week to check on the progress of experiments. "Take a seat," Bill said rather laconically, and then added "So, where are we on this protein rhabdosin? How are the experiments going?"

Sam had already pre-empted the questions and rehearsed the answers early in the morning. "I have already placed the order for the carcinogen and it is expected any day now. I can administer it to the first group of mice the moment I get it. The other 10 mice are already under regular treatment of rhabdosin, so that a later administration of carcinogens, hopefully, will not induce cancer." Sam said rather sheepishly.

To this, Bill started to look more acutely at his coffee mug, avoiding a gaze at him. This meant there was a problem. Bill avoided eye contact when he was about to say something grave. "Well, you know what? I am not too keen to continue funding this project if I don't get any conclusive direction from our project by the end of next quarter. You see, I have to answer to some drug companies. I set expectations for them, and my company's name is at stake." This was May. Next quarter end means September. A little over four months. That's all he had.

"Bill, I ..."

Bill cut short his dialogue by gesturing his hand. "I admire your brilliance, Sam. But brilliance does not keep our company running." He brought his face closer to him by leaning in from his chair, "Results on time," he almost whispered to Sam. Whispering meant he was literally starting to lose his temper. This Sam knew. "That's what keeps these experiments running! I will have to scrap this project if I don't see anything conclusive."

While walking out of Bill's room, Sam paused at the cafeteria. He realized that he had to get started right away. He felt sorry for the 20 mice who, like every other innocent creature on this planet, knew nothing of man's crooked scheme of things in which they were just a small piece, but a crucial piece nevertheless.

He had two groups of mice. Ten of them were being treated with carcinogens. The other ten would receive rhabdosin initially and then the carcinogen.

He called them Martyr 1, Martyr 2, Martyr 3 and so on. In short, he called them as Martys. He had developed a special connection towards them. What if rhabdosin did not prevent or cure them after they had cancer from the carcinogens? What if they had some other diseases as a result of carcinogens and rhabdosin? What if? What if?

He remembered the teachings of Gita where the big picture and larger benefit for humanity may warrant small sacrifices. The lovely face of Erin flashed in his mind for a while, her eyes were ocean blue and steady yet there was something feisty in her diction. Mice, Gita, Erin, rhabdosin, all created a little vortex in his mind and all he knew was that Marty 1 to Marty 20 were going to die if his hypothesis on rhabdosin was not really well founded. He decided to give his every ounce of effort to ensure that Martys injected with rhabdosin came back to normalcy. It is not only Marty's recovery but an allegorical recovery of millions in a way. A path to be paved for the posterity.

After six weeks, the carcinogen-treated mice had lesions and small tumors. In the ones injected with rhabdosin, the lesions and tumors shrank. The ones which had been administered rhabdosin beforehand had very few lesions and never developed tumors. None of the mice administered rhabdosin showed any signs of toxicity or abnormality. Sam had his proof! He was ready to go forward with more experiments. He hired a contract research organization to conduct ADMET (absorption, distribution, metabolism, excretion, toxicity) studies with rhabdosin, which indicated very favorable results with good serum stability and very little immunogenicity and toxicity in animals.

His project was funded. With permission from his university and institutional research board, as well as money from the sponsors, he moved forward to applying to the regulatory agency with an investigational new drug application to try rhabdosin in patients with late stage cancer.

six

Erin's crisis continues

Erin came away with the feeling that she had absolutely no control over her body any more. At first it was nature that had conspired to play havoc with her cells. Now it was the strange feeling she had for the research scientist Sam who was trying to convince her that he could conquer cancer one day. In his online lectures, which Erin watched once in a while, he also talked about collaborating with a gene therapist to treat breast cancer and have patients whose genes may be altered for the rest of their lives. Was this a crazy sci-fi super fiction! Added to this unreal brew was her rather real liking for this researcher cum poet. She could not ignore the admiration and awe she felt for this brilliant committed man, who had the brains and the magic to grapple with one of the most powerful of diseases. Nor could she deny her attraction for his sensitivity and wit, to empathize and connect with the diseased. She wished she wouldn't be so scared of his ingenuity, but try as she might, her debacle-ridden past, coupled with the fact that Sam was so different from her in very many ways, made her very uneasy and wary of him. An errant thought gnawed at her brain like a persistent bug. Would Sam try to be so friendly with her had she not been hale and hearty? Would he have valued Erin Goldberg, a genuine human being, a respected librarian, doctor's secretary and a self titled poet, if he knew she was an invalid, diseased and handicapped? Would he just shower her with pity, and worse still, brainwash her into being useful to him for his research?

Thankfully Erin was very busy over the next few months with summer plans of extending the library computer rooms and other online facilities. She did not spare herself and scheduled her days in such a way that she'd have little time to think. Dr. Lee had gone on vacation to China, so she didn't have to deal with his many well-meaning reminders to come in for physical check-ups. They were making her feel that her days were numbered. While on vacation he had asked his full time secretary and Erin to take care of his office matters. A junior doctor from another group was attending to his regular patients. She tried hard to prevent her mind from going back to that early spring day when she had made such a fool of herself in front of Sam. It was odd it mattered so much to her how she had appeared in front of a guy so different from her in cultural background and profession. She would have preferred him to remember her only as Bonita, the poet. She thought a lot about what Sam had told her about his childhood in India and how he had aspired to overcome geographical boundaries, to realize his dream of a healthy world. She was not sure she understood everything about his loyalty to his roots or remembered much of the technical jargon of what he was researching, and its future results. But she couldn't wipe out a strange feeling that he had instilled in her that afternoon in her library office. It was almost like a faith that should she need him, he would be there for her. That was more than she could even say of her long-term acquaintances. But that was not enough for Erin. Because she had a genuine liking for him, she wanted to meet Sam on equal terms in mind and body and because she knew she couldn't, she did not return any of Sam's messages. She somehow resented the fact that Dr. Lee and Sam, though they were in separate fields, seemed to be super anxious about her. Why were they always thinking the worst of her body? It made her somehow feel inadequate and apprehensive of her future.

Around fall Erin was in the shower doing her usual breast self examination, when one morning she suddenly felt a lump just to the side of her left arm pit. For a few days she ignored it, but when she told Dr. Lee he immediately told her to get a biopsy done. The biopsy proved benign, but the lump didn't go away either. In fact it seemed to be growing. Erin was worried. Sometimes she would wake up in the middle of the night in cold sweat, her heart beating fast, as it seemed that her lump was consuming her body. Finally she couldn't sleep at all, panicking that while her mind was at rest, the lump would kill her body and she would die without knowing it. When Dr. Lee returned, she told him that she would opt for double mastectomy like her aunt. But Dr. Lee pointed out that since she did not have any malignant tumors, the True Pink organization would not take up her case, and when she called them up, that's just what their answer was. Dr. Lee noticed how distraught she was becoming, and suggested she go on a vacation to ease her mind. He also said she could ask Sam if his research could give her some more information on her options. But Erin's personal feelings for Sam got in the way.

One day, Erin decided that she should have that lunch date after all. She had a plan. She would meet him in the Indian restaurant near the Aurum University library after work.

seven

Sam's memories

Sam remembered the days when he first met Erin. Wasn't it weird that he had felt so comfortable sharing his thoughts and work-matters with a girl from a totally different background? A girl he had just met? In fact he had started to look forward to the Thursday evenings, just for the sake of looking at Erin and being able to talk to her. Although he respected her outlook and enjoyed her company, he knew there was not much erudition in Erin, yet there was this indomitable urge to learn more, to imbibe cultures and religions across the world. In fact when he asked her where she had heard of the *Gita*, and how she had known all those details, she had replied "My friend gave me that book as a gift after my mom died. It gave me a lot of solace."

Erin had started to bring nice snacks every Thursday evening for Sam, since Sam used to be dead hungry after his day-long research and also because he sped to the library right after work and hated store-bought packaged food. The snacks Erin made were mostly baked and sometimes could do with an extra dash of condiments, but whenever Sam appreciatively gobbled the pumpkin pie, or apple strudel, or the blueberry and walnut cake, her eyes lit up. And this color in her eyes made Sam feel a bit puzzled, but also at home in a way. It reminded him of his mother who used to sit with him through all his meals until the last few weeks before taking to bed. In fact, on her orders, Sam had to go and have his meals in front of her bed just so she could see him eat. It was an unusual connection, but a connection nevertheless.

Then, his experiments and the demands of his pursuit totally eclipsed everything else. Many times in the course of his busy schedule Sam had wanted to call or text Erin. Her down to earth, earnest remarks had somehow become important to him. He kept thinking she would get in touch soon. Was it his imagination or did her eyes really turn from dark grey blue to a sunny turquoise when she saw him? Well, only time would tell.

Several months went by. His rhabdosin project was going well while he waited for the investigational new drug approval. Then one day he had a patient named Kathy, in advanced stage of ovarian cancer. Apparently, she was Erin's cousin. Sam decided to visit Erin in the library again after a long time. Erin had appeared to be distraught that day. He invited her out for lunch, but she didn't feel like going out. They decided to make another date.

One afternoon Erin texted him. She had something very important to confide in him. "Could you please meet me after work? How about the Indian eatery near the library? Need to discuss urgent stuff."

Sam's brain immediately started working in overdrive. What kind of urgency? Problems in job? Did she get fired at Dr. Lee's clinic? Not a problem, he could handle that. Did she have anyone bothering her at the library? He could ask her to quit the library job and convince her easily that she'd make more money if she joined his friend's clinic not far away from her home. That was easy too. Was some friend of hers having some issues? Or was it something else altogether? Did she know how much he admired the way she had made a good life for herself after her mother passed away without leaning on anybody else but with her own abilities, strength of character and wits? All this touched Sam's heart so much that when he found out at the library that Bonita was actually Erin's pen name, that very night he had penned a poem in response to hers. He wanted to remove her pain for good, to be close to her. Could she be calling him to dismiss him from her life, as a reaction to something he had said or done at the library when he last met her to talk about Kathy? Well, he had just tried to console her for she had really seemed very upset about something. Whatever it was Erin surely was keeping him guessing and on his toes! Sam kind of enjoyed that too in a way! Didn't that make life a bit more interesting? After all, what's the fun in winning a lady without any challenges? Sam saw infinite possibilities and couldn't suppress a smile at the prospect of being able to be of some help to Erin.

He again asked himself the same question he had asked ever since his last meeting with Erin "Was he in love?" As an answer he could not come up with a vehement no, because the poem that lay in his saved folder of emails told him otherwise. He would send it to Erin when the time was right.

Sam arrived at that Indian restaurant a good half hour earlier than required. He was still puzzling over how he had managed to wrap up his experiments so soon that day. He ordered a cold beer and decided to wait for Erin. He texted her again.

eight
Erin reaches out to Sam

And there she appeared, almost like a genie of Alladdin's story. She looked a bit fatigued and tired. But her usual casual elegance and easy smile made his heart skip a beat. Sam quickly got up and pulled up a chair for her. "What's going on, Erin? Is there something bothering you?"

"Hi Sam!" said Erin and sat down rather heavily on the chair Sam had offered. She took a deep breath. "I'm sorry to spill this awful news to you so abruptly like this! But I heard this a few days ago, and it hit me like a bolt of lightning. It is about a dear friend of mine who means a lot to me." She paused and swallowed hard. "But before I share it, will you promise to keep this confidential?"

"Oh! come on! You can trust me, Erin!" He replied almost instantly but he had to admit he was a tad apprehensive ... was this her lover that she was talking of? Well, whatever it was he would surely listen, he could not see this lady suffer for any reason!

Erin continued without looking at Sam, just staring at the glass of water she had taken a drink from. "I know it's going to change my friend's life. About a month ago, she thought she found a lump in her breast. She was kind of paranoid. On her mother's side of the family, most of the women and even some men have had breast cancer. Her mother had died of breast cancer just like my mom. My cousin Kathy, who is only 28 and is your patient, has ovarian cancer. The funny thing is that she is also related to Kathy. So possibly this is something that runs in the family. Her aunt Linda had to have double mastectomy. Of course, you know about Kathy's history."

Sam sat bolt upright and asked "So what happened? Did she get a biopsy?"

"Yes, it was negative. She was overjoyed. But her doctor said, not so fast! When he heard her family history of breast cancer he sent her blood to some lab for a DNA check. It has come back positive. He said she has two different mutations and there is an 80% chance that she will get cancer—maybe breast cancer—maybe ovarian cancer. What are these mutations, do you know?"

Sam pondered for a while before answering "I don't know exactly what your friend has, but she probably has mutations in the BRCA1 and BRCA2 genes. They are short names for breast cancer susceptibility genes. Remember, we talked about them before? They were discovered in 1995. They are also called tumor suppressor genes. Normally they repair the DNA. When DNA breaks down, these BRCA genes, actually proteins made from them, will fix them. When these BRCA genes are slightly different from normal, say when they are broken or a chunk has been taken out, they don't work so well. These changes are called mutations. To get cancer, a lot of genes have to be out of whack. Your friend probably has the BRCA mutations.

"Maybe these guys made a mistake in checking the DNA. Can we go to another place to check her DNA?"

"I am afraid not. There is only this one lab, called Myriad Genetics. Only they screen for these breast cancer gene mutations, because they hold the patent. However, for poor women or women without insurance, they reduce the price considerably. How much did it cost her?"

"I don't know. I am told that her insurance will cover it. Can you explain how she can inherit cancer because of the mutations?"

"We are humans. All our genetic material is in DNA, or deoxyribonucleic acids, is distributed in 46 chromosomes. These 46 chromosomes are in each of our cells, 23 from mom and 23 from dad. Only germ cells like sperms have 23 chromosomes. The 23 that you got from your mom, one of them, number 17, has BRCA1 gene and the other, number 13, has BRCA2. Only, your friend's genes aren't normal. They are kind of broken or mutated. If her DNA is damaged, they won't be fixed easily. Of course, other factors are responsible, too. Like what you eat, whether you sunbathe, whether you smoke. An American diet with a lot of fat is supposed to increase cancer. It has been well-documented that when women from Japan come to the good old USA, their cancer rate increases. In fact, more and more young women are developing breast cancer now."

"How are these mutations created?"

"Different reasons. Bad habits, ultraviolet rays, getting old. The bottom line is that somehow our DNA changes. It may get a bit more technical but I will try to explain you as simply as I can."

Sam quickly walked to the sugar rack in the coffee shop and brought a handful of sugar packets of different colors. Four colors to be precise, a few brown sugar packets, a few pink Splenda packets, few blue Equal packets and few white regular sugar packets. Then he spread them on the table and formed a string with them with a sequence of a brown, pink, blue and white packets. Then he went on, "DNA is like a long string like this which has instructions for your life. This language has only 4 letters in the alphabet. Imagine each of these packets as the letters in an alphabet. They are nucleotides called A, T, G, C. The order of these nucleotides is called the DNA sequence. When a cell divides, the DNA also divides. Sometimes there is a mistake, like a typo, and one nucleotide changes. Sometimes, the Sun's rays will just break the DNA into pieces. Sometimes, some chemicals in the food will change your DNA. These chemicals will get tangled with DNA and won't allow it to replicate properly. If you don't have the repairing proteins, then you may get cancer."

"Will these two mutations alone cause cancer?"

"Cancer is normally an accumulation of many mutations. Let's say from 5 to 100. Some mutations are critical for causing cancer. They are dubbed driver mutations. Other mutations make the cancer grow faster. They are called passenger mutations. As you age, these mutations or errors accumulate. If you smoke or lay under the sun too much, there are more mutations. Sometimes, may be you didn't do any of these things, you just inherited these mutations from your mother or father, like your friend."

"So that means that if she has these mutant genes, when she gets married and has children, her children will get these bad genes, too?"

"Not necessarily. They could get the good BRCA genes from her husband."

"She is a very close friend. I feel that she is like a sitting duck. She is just waiting for a killer to attack her. Where can she go? She can't change her body."

Erin started to sob. Sam held her hand to comfort her. It agonized him to see her in pain. He held out a paper napkin for her. Something screamed in him to comfort her.

Erin rubbed her nose, recovering a little. She was embarrassed, but she tried to hide it. She asked defiantly, "So, what are you scientists doing about it? I know that you are supposed to be the expert."

"Actually, we can do something to prevent cancer. You, my dear, are talking to the right guy! We are thinking about a revolutionary idea. As you know, we have discovered a bacterial protein called rhabdosin, which can cure cancer in mice. We are applying for regulatory approval of a trial where we will be injecting rhabdosin in people with cancer. Your cousin Kathy will be in that group. Your friend cannot join that group now, because she does not have cancer yet. She can join another trial, which we call somatic gene therapy. We are not ready yet, but when we are, the gene for rhabdosin will be introduced in the chromosomes of her hematopoietic stem cells. She will be making her own rhabdosin, which will enter her cancer cells, if and when they appear in her body. I am going to talk to a well-known gene therapist who conducts such trials."

"Is your approach different? What have you done that has made you famous?"

Sam's eyes sparkled as he explained. "Let me tell you how anticancer drugs are made. There are genes in your body that regulate the growth of cells. Cells go through a cycle of events and some genes control the growth at some check points. When there are mutations in these controlling genes, the cell starts to grow rapidly among its non-growing neighbors, giving rise to a lump of fast-growing cells called a tumor. Then, there is a dangerous process called metastasis, when the tumor cell moves from one tissue to another; that's when it becomes very difficult to treat the cancer."

"So, how do you stop this runaway growth?" Erin was listening with wide-open eyes. She actually looked terrified.

"The drug makers normally fixate on a single or limited number of these critical steps of regulation called targets. Then they use thousands of chemical compounds to look for a few that will inhibit these targets and slow down the growth of these tumors. These inhibitory compounds are called hits. They are then tested first on animal models and then on a few patients to see if they are toxic. Only drugs that are nontoxic, but still have cancer-regressing activity will be approved for marketing by the FDA. As you can imagine, only a few pass through regulatory approval and it could cost up to a billion dollars for a single anticancer drug to reach the market.

"Well the cost is certainly high, but at least the cancer is kept in check forever, right?"

"You wish! These drugs only stop one or two steps in cancer growth. The cancer cell can change its path or it can just pump the drug out and thus become resistant to the drug. There is also one more problem. The targets of cancer cells are also present in all normal cells, but there are some normal cells that grow rapidly, like hair follicles or cells that line the stomach. So, after a few weeks of these drugs, these cells die, too, causing hair loss, nausea, diarrhea, etc., in the patient. Sometimes, the side effects are worse than the cancer itself and the patient cannot stand it. In such cases, patients refuse treatment."

"Sam, I am dying to know about your drug. I know you told me in the coffee house before, when you were working with mice, but I forgot. Tell me again."

"And I'm dying to tell you. But neither of us has to die yet. This drug will save us! When I was in India, working in the field of infectious disease, I made an interesting observation. Bacteria can protect us. There are some bacteria that form thin films of growth on the tissues of some type of patients. They could live there for years. They consider the human body as their home. Other scientists, as well as people in my research group, have shown that bacteria can use proteins as their weapons to attack invaders like cancer. Bacteria have used billions of years of evolutionary wisdom to design these proteins as weapons. Rhabdosin is one of those proteins that we are going to use on Kathy, if or when we get regulatory approval. For your friend, we might also be able to put in the rhabdosin gene in her genome some day."

"Are you serious? It sounds too good to be true. Won't there be some bad effects when you introduce DNA of these smart bacteria into human beings? Aren't bacteria foreign to us?"

"Well, yes and no. Bacteria are not really foreign to us. Bacteria were the first living creatures to arrive on earth about 3 billion years ago. Millions of years ago they entered lower forms of life and became trapped in them. As the theory goes, you know, two billion years ago, bacteria got trapped in ancestral eukaryotic cells to help remove toxic concentrations of oxygen and started a symbiotic relationship. This was called endosymbiosis. After hundreds of millions of years of evolution, these bacteria lost most of their genes to the host, and therefore their independent identity. They became part of the host cells, named mitochondria. Having one more bacterial gene such as rhabdosin in our nucleus, although unheard of now, may not be of great evolutionary significance. Of course, no one knows what exactly will happen after gene therapy, but I am very hopeful it will work."

"The whole thing sounds like science fiction. You don't seem like the kind of man who would take advantage of a woman in a vulnerable position. I will go home, sleep on your fantastic idea, and dream about a world where there is no cancer. I will also speak to my friend tonight." Erin felt a little calmer than before.

Sam tried to console her. "Please hold on to all these positive thoughts. After all, there is a 20% chance that she will be the victor in this battle."

"Well, I'll try and explain all this to her. And will you talk to this gene therapist as soon as possible?"

"Yes, I will. Dr. Warten lives in Washington D.C. I have not met him yet, but I have heard a lot about him. But I also need to talk to a patent lawyer.

"Why do you need a patent lawyer?"

"It is a kind of government sanction that you have to get when you invent a new procedure and want to protect it from copying by others. This is why Myriad Genetics is able to test people for these gene mutations and no one else can do it without their approval. They discovered those two genes BRCA1 and BRCA2 and the cancer causing mutations in them that your friend probably has. Everybody has these genes, but when they are defective, women or even men can get breast cancer. Men can also get prostate cancer instead of ovarian cancer, but rather infrequently."

"I am not too sure that she will want to undergo this therapy, but I will ask her to think about it. What about her future kids? Will they be cured, too?"

"Erin, one step at a time, please. That is called germ line therapy. We are not there yet with human beings, but we can insert heritable changes in mice. We should be able to figure out how to do it. Let's not worry about it now. You haven't even noticed I ordered sizzling Tandoori chicken for you from the Indian barbeque pit. As far as I remember you told me you always wanted to try the red gold grilled curry chicken. See I didn't forget!" Sam pushed her plate closer to her.

Erin and Sam both ate silently. Sam seemed to be enjoying his non-veg thali. Erin couldn't taste anything except the spice and the heat. She looked at Sam in between bites of her Indian rice and chicken and said to please him, "Wow! it's really got flavor and a sharp taste. Quite … quite different!" she ended lamely wondering if she had offended Sam by not taking an instant liking to Indian cuisine. But he accepted her comments and continued finishing his helping with a healthy appetite. When he bent over his plate he was so close she could touch his head of vibrant dark hair. Actually she wanted to reach out and touch him. Yes, she did have some genuine feelings for this earnest man, otherwise why would she bother to spin a story about her friend? Was she afraid that if he knew it was she who had the mutations he would start viewing her as being somehow less worthy of being his friend? From whatever she knew of Sam, she knew he probably would never demean her, but what about pity? She knew she couldn't take that from any man she truly loved. They would have to respect each other as equals in mind and body. Erin could never be happy with less. And she wouldn't want Sam to be either.

The rest of the meal went well, but rather faster than Erin wanted. When they said goodbye, Erin brashly hoped Sam would hug her again as he had at the library. But he didn't. He did touch her lightly on the shoulder though to say goodbye.

"I'll be in touch," he said waving goodbye. These shy Indians! Erin tried her best to control her emotions. "Who knows how long it would be before I see him again!" she thought, stemming back tears that were about to cloud her vision. As she got into her car and drove off, she had a premonition that soon she would have to take an important turn in her life.

Sam and Erin had had their first intimate talk, even though she had lied. Was it a white lie? Was it a lie about love or for love? So far as Sam was concerned a fleeting question had crossed his mind. He did not consider himself to be a very good judge of character. But he had an uneasy hunch that Erin was concealing something. Could it be that she was too reserved and withdrawn to divulge it was actually she herself who was the real patient? Maybe she didn't feel ready to open up to him yet? Oh God! How desperately he hoped in his heart of hearts that it was not a lie. Let it not be Erin! He prayed silently, instinctively, unconsciously. Let it be anyone else in this world but her! And he even forgot to admonish himself for being selfish!

When Sam went home, there was a message waiting for him. It was from George Rivera, a patent lawyer.

nine
George, the patent attorney

George was running late for the train. In the surreal darkness of mid December, George was sprinting through the packed parking lot with the heavy backpack loaded with briefs. He had to park at the farthest end of the lot. Without the layers of clothing and the briefcase he would have been fine with running, but not today.

A great anger was consuming him. Alicia spent another three hundred dollars to get her hair done yesterday. That was five hundred dollars on hair styles alone this month. Last night's showdown was becoming a routine. Alicia was climbing the stairs. The light from the great chandelier hanging over the foyer glinted off her perfectly done hair. He noticed that something was different.

"Is that a new style?" George asked.

"Yes, do you like it?" From the top of the stairs, Alicia turned toward him and stood with an alluring pose. She did look gorgeous.

"What was wrong with the style you got last week?" he asked.

"Men. They don't care! You expect me to look like a woman from the barrio!" she said. There was a not-so-hidden jab about his childhood.

"Don't you understand that we do not have that kind of money?" His voice rose. It sounded sharp in his ears.

"Don't you yell at me!" She turned around facing him with her palms firmly planted on her waist. Her voice rose contorting to a screech. George felt the anger throbbing in his head. Yet she looked like a goddess in her sheer nightgown. This was the woman he fell in love with eight years ago. He almost felt like embracing her in his arm and shaking her wildly to make her understand that she could not continue spending money. Perhaps, with the feel of her body next to him he would calm down and they would make up. Then he noted that he had not seen the nightgown before. Another hundred dollars, perhaps! George thought. Going anywhere near her at this moment would be futile. He could see the luxurious bed through the open door of the bedroom. No space in that bed. He grabbed his favorite pillow and rushed off to the office downstairs.

What did she expect? He earned a decent amount of money with the usual patent stuff, a machine configuration here, a new kind of gizmo that somebody figured out there. She could have a good life being the stay-at-home mom she was, if she was reasonable. Her life revolved around being popular. There were gym classes, hanging out with friends and other such things. Looking fantastic was a big part of it. She loved Katie, their five year old daughter. Again, it was all about being social. Even at this young age, Katie knew about designer outfits. There were those crazy American Girl dolls. It seemed she already had half a dozen of those insanely expensive dolls complemented by their fancy wardrobes. In the meantime, his assets were dwindling fast. With the second wave of housing crisis within a decade, his fancy home was significantly under water.

The train was entering the station, but George was still in the tunnel. He tried to shake off last night's drama from his head. However, as he climbed the stairs to the platform, the express train left. The next train was not an express one. He had to wait for the train after. That meant he would be late by fifteen minutes. He would have to miss the first appointment with a client. Fortunately, she was not a big corporate client, but a small business owner who wanted to do a simple process patent.

Once inside the station room, George dumped his backpack on the bench, fishing out his cell phone from his coat pocket.

"Hi Susan, I am sorry. I just missed my train. I will be half an hour late. Is that OK?" He expected an angry customer voice from the other end.

"That is OK. I am running late too. I will talk to you soon." Her voice was barely audible. The hum of her car speaker phone—must be a cheap car and a cheap phone—was drowning out her voice. Just like his decade-old Honda. Of course, Alicia drove the latest model Jaguar. She was angling for a set of filthy big diamond earrings for the next Valentine's Day.

"I should get Alicia out of my mind. I really need to find a huge potential client. Need to do more research. Everybody is getting onto the gene patent bandwagon. I should look into it."

Once inside the next train, George remembered the news snippet on CNN about Myriad Genetics gene testing and their patent. He punched in the terms in Google on his cell phone. However, there was no signal. The train had already entered the desolate industrial region at the periphery of the city. He decided that it would be best to use the rest of the time in the train on Susan's brief.

ten

George and his aspirations

A fire was burning in George's belly. After last week's rough start with Susan, the follow-up meeting with her went well, but today's mid-morning meeting with his boss did not. His boss was hinting that he must snag a high-profile patent job, preferably with a corporate client, soon. It was clear that his position as a junior attorney would be in jeopardy otherwise. It was almost lunch-time. He needed to clear his head and got ready for his lunchtime run. When he passed the corridor he noticed that the sun was shining outside. The weatherman predicted that the temperature would be mild today. No snow or wind. His foul mood changed for the better. Outside, he took his usual route. A Bach fugue was pouring in his ear from the iPod. As he concentrated on the twists and turns of the fugue, he calmed down. Within ten minutes, he no longer needed to think about the music.

Instead, he thought about the recent Myriad gene screening controversy. He only had a short time to do any real research about it since last week. Although the patent was awarded a while ago, it had been a controversial one ever since. Advocacy organizations were putting enormous pressures on the legal system against gene patents. There was a tentative provision to get this kind of genetic screening covered under the Affordable Care Act, but with the election approaching fast, it was under great scrutiny from both parties. It seemed that the whole Affordable Care Act might be repealed. The bigger issue was that with the newest Myriad screening patent. It would be a great dilemma for a patent attorney because it was a classic chicken and egg story. Without patents no venture capitalist would invest a penny in helping bring to market such an innovation. The idea would die at inception. Sick women wouldn't be able to get the right treatment. Since the days in law school when he decided that he would help American progress through patent laws, he believed in patenting, but could not imagine ways out of this dilemma.

However, how could he ignore what these advocacy groups were saying? It was true that if the Affordable Care Act was repealed, Myriad might charge an enormous amount of money for the screening making it only affordable to the richest. If it stayed, then men and women would have to depend only on Myriad Genetics or its licensees until the patents expired. If Myriad's testing was wrong, then what would happen?

On the other hand, assuming that the test results were correct, then what? There was no known surefire treatment for the breast and ovarian cancer, feared to be due to BRCA 1 and BRCA2 mutations. He should try to figure it out what was happening in that area, but how? He slowed down before the traffic red light and checked his watch. It was time to turn around and go home.

"Duh, there must be something brewing in the universities around me. I must check it out as soon as I reach my office." George was anxious to get back to his office and took a shortcut to reach his building. He passed the graceful St. Mary's of Angel, a Catholic church. He was raised by devout Catholics, but, like many others, strayed away from the faith, particularly after all the sex abuse scandals. Fortunately, he never had a bad experience in his church.

As he took the elevator to his 16th floor office, he remembered the Monsanto case before the Supreme Court about their patent infringement lawsuits. If Monsanto could manipulate soybean plant genes, somebody must be doing some work on human gene manipulations. Monsanto had modified the soybean genes so that the plants were not affected by a weed killer. Would there be a modification that could be made to prevent cancer? It was odd that he did not hear anything about it, but he hadn't looked into medical patenting. His firm never took initiatives in that field. Maybe that was his ticket out of the stagnation. If he could take a lead in genetic treatment patents, it might change his life.

As soon as he entered the office after a quick shower and lunch, he forgot to follow-up on genetics. There was no time to think about it. First he dealt with the voicemails, calling back clients with bruised egos. There were emails. Revising two long patent filings took up the rest of the afternoon. They were about simple mechanical devices. By the time he completed the day's tasks it was almost seven, long past his scheduled time to go home. He toyed with the idea of staying at work longer to research possible gene manipulation for cancer cure. However, his boss did not encourage people to stay in the office later than 8 pm. Perhaps it was a policy from the building security management. His boss had no problem when his employees had to work from home for several hours. George dreaded going home to face Alicia. She was on a warpath since last week. He had not slept in his own bed since then.

At home, George found Katie playing with the baby sitter from the local university. The sitter informed him that Alicia had gone to a fundraising planning meeting that evening. That was another of her hobbies. Spending time with cool people at the fundraising dinner was a sure way to be popular. However, George's bank account got lighter every time she attended one of those events. Katie was pleased to see him and gave him a big hug. He picked her up, but she wiggled out of his grasp. Within seconds, she went back to her American dolls and make-believes. He was relieved that he did not have to face Alicia.

George checked out the refrigerator for some food. He was famished. There was nothing substantial inside this sparkling clean refrigerator, but there were several Lean Cuisine packages in the freezer. He pulled out two old ice-crusted ones from the back hoping he would not make Alicia mad by eating her favorites. There were several beers. After sticking the Lean Cuisine in the microwave, he sat down at the kitchen table with a beer.

The immaculate and hardly used chef quality kitchen gleamed around him, but Myriad screening research was firmly planted in his mind. He finished his meager dinner as soon as possible. Not letting the Lean Cuisines cool was a bad idea as it burned his lower lip. With the beer in hand, he entered his den, his sanctuary, at the back side of the house overlooking the manicured lawn. He dove into his regular research tools. It seemed that there were not many gene-related patents in the pipeline. The Myriad patent was the predominant one. Curiously enough, a large amount of information related to Monsanto's Roundup Ready patent controversy was popping up frequently. George was intrigued and delved further into the gene research. It was time to do serious scanning of the conferences related to genetics. There must be somebody in the universities around town who would know more about the subject. Gene therapy showed up several times in his searches.

He remembered the celebrated case of *Diamond v. Chakrabarty* that every patent law student had to study. The first life patent assigned to GE about the oil-eating bug was approved by the US Supreme Court. Unlike the human gene screening, this microbiology related patent was about the patenting of living things, in this case a bacterium.

As he skimmed the abstracts of an upcoming meeting in Washington D.C., it dawned on him that there was a large contingent of scientists from Chicago who were attending that conference. The abstracts were quite obtuse, yet some of the folks from Chicago might be able to shed some light on the recent innovations in this field. If he was lucky, they might need some legal help as well. One of the author's names, a Dr. Zazac, seemed very familiar to George. Did he meet him somewhere a couple of years ago? He typed his name in the LinkedIn people box, but no luck. He must have his contact somewhere in his office. He had to wait until next day to dig up the contact information about this Dr. Zazac.

"I am leaving now. I have put Katie in bed. Mrs. Rivera already paid me," the baby-sitter appeared at the door. She was lugging a heavy backpack—must have been doing her homework while watching Katie. George felt a sense of respect toward her, remembering the backbreaking yard work he did as a teenager for wealthy clients around Albuquerque. His dad was often away from home, visiting free medical centers near Indian reservations where he treated poor people. His mother, a capable woman until her sudden death, never complained about his absence or the lack of money. She was happy that unlike many people around her, her family had legal status. George was determined to get his citizenship papers when he turned eighteen. That was his first encounter with the US legal system, and not a pleasant one.

"Your father's good deeds would bring many blessings to you someday. You will see." Whenever George complained about not being able to afford something, Mrs. Rivera reminded him in her cute accented English, instead of her usual Mexican words. It seemed to George that mama used English to emphasize. He missed his mama, particularly her cooking.

After the baby-sitter left, George moved to the old sofa with his notebook. He continued to dig deep into the scientific work of Dr. Zazac and his colleagues at the university, reading everything he could find on their genetics research. Most of the reading materials were completely unfamiliar to him. However, being used to doing law research, he managed to break down the basic facts in small chunks, taking copious notes. Surprisingly, there were only a handful of patent applications in this area.

It was getting late. There was no sign of Alicia yet. He knew that Alicia was a popular, but faithful, woman. He could not deny that her popularity and her family connections helped him a lot to establish his career. It was strange how she managed to dig up the right connections for him. Her socializing certainly helped him.

He wrapped himself with the blanket and dozed off.

As he was dozing off, George recalled his early years as a lawyer. After graduating from a well-known law school but with a last name like Rivera, he was not sure how he would be able to find a position in this big town. It was basically Alicia's connections that worked. He was still in law school when he met her.

Meeting Alicia at the Hilton hotel bar was truly fortuitous for George. Sitting on a high bar seat, the tall and beautiful Alicia was towering over her girlfriends. Her slender manicured fingers enhanced by green nail polish were wrapped around a green Margarita glass. She must have been wearing a green dress as well. He could not remember. She was attending the wedding of one of her friends from college. There he was, bar hopping with his group of friends after the bachelor party of Kenny, a close friend from the engineering college. He was first attracted by her tingling laughter that seemed to spill over the festive crowd like foil confetti falling from the sky. Green nails and laughter—that's all he remembered. Does one need to notice anything else? He fell in love with her immediately. With great intensity, he chased her for a year. He found an engineering job. She would not pay any attention to him until he joined the law school. Entertaining her was expensive with the burden of the student loan. However, he could not stay away from Alicia. They broke up many times, but he had to get her back.

Once he graduated, Alicia moved in with him. She turned on her charms like an open faucet to get him a job at the small law firm that her extended family used for their businesses. The law firm was trying to get into the lucrative field of intellectual property law. Without any connections in Chicago, George did not think he could do any better. With his engineering background, he had decent success in rounding up business for the first few years. It was clear that Alicia required high-maintenance. With his mother long gone and his father entering the shadows of dementia, George drifted away from his only sister. He knew that she would not approve of Alicia.

With her biological clock ticking away, Alicia and her family kept pressuring him to settle down. On top of the student loan, he had to borrow a lot to buy the twenty thousand dollar diamond engagement ring. Alicia married him. George would never forget the day Katie was born. He knew that it was worth loving Alicia. Within an hour of giving birth, Alicia looked perfect. She was writing "thank you" notes to everybody of some importance, tossing aside cards from people from the wrong side of the tracks.

eleven
Alicia helps George

Alicia's Manolo Blahnik heels were clicking away toward the staircase. He held his breath—would she or wouldn't she let him sleep in his own bed tonight? His notebook was already in sleep mode. Perhaps it was a good time to make a move. He got up to meet her by the foyer.

"How was your dinner, honey? Hope you had a great time at the party," he hugged her gently, planting a kiss on her perfectly made up left cheek. There was a faint breath of alcohol that delighted him slightly. Knowing how careful Alicia was about drinking and driving, he concluded that somebody must have dropped her off at the house. She was usually in euphoria with a little alcohol in her system. This was a good time to patch things up with her.

"Let me take off your boots," he offered. Alicia sat down on the fourth step of the staircase and extended her long slender left leg toward him. For a moment she looked very tired. As he pulled off the other boot, he could not help but admire the solid quality of this apparently delicate object. These Italians know craftsmanship. Every part of this shoe was probably protected by some kind of intellectual property law. It was no wonder that Italians were the first to come up with the concept of patenting and, of course, Greeks had to be a part of it too. Italians in the Greek city of Sybaris, now in southern Italy, figured out how to patent cooked food. From then on, mankind patented every imaginable object and processes. The downside was that now poor folks like George had to shell out big money to keep up with the status of their wives whether they could afford it or not.

As soon as he pulled out the other boot, Alicia removed her delicately sparkled panty hose. Sitting on the second step, George began massaging her feet as Alicia leaned back on the steps, half-closing her eyes.

"What are you smirking about?" Alicia's voice jerked George out of his train of thoughts. He was thinking about how to secure his job and make big money to buy many more Manolos, American Dolls, expensive cars and diamonds. This million and half dollar home might keep Alicia happy for a while, but who knows? Right now, George felt that he could do anything for this gorgeous woman with a fantastic hairstyle. Alicia was an independent woman. He did not have to tag along with her to every event. Yet, she returned home to him and Katie with some useful information and connections. He felt a tinge of guilt about squabbling over her salon expenses.

"Oh, nothing! You know what is on my mind when I massage your feet! How was the fundraiser?"

"Oh, yeah! Don't I know that!" Alicia continued, "You should have been at the party!"

"Why's that?"

"I met this big shot MD/Ph.D. guy called Tom Zazac. Gina introduced me to him as a wife of a big shot patent attorney Mr. Rivera. He remembered you from a patent law workshop you led a while ago. I suppose he was made to attend it. It seems you made quite an impression on him. He does not seem to be the kind of a person who would willingly attend a legal workshop."

"Really? I came across his name this evening. I need to get in touch with him. You are awesome, Alicia." He held her tight and kissed her. This was serendipity. Only an hour ago, he was puzzling over how to get in touch with somebody like Dr. Zazac. There was no doubt that Alicia brought good luck to him.

twelve

Getting to know the scientists

The wonderful cooking scent from the new fusion restaurant Cayenne was making George hungry. He was standing in front of the restaurant, as he had never met Dr. Sam Roy and had forgotten how Tom Zazac looked. He was not sure that the waiters would remember to bring them to his table. It was 12:30 pm. His guests Tom and Sam were late, but he was used to being patient, particularly with potential clients. Tom was a difficult man to get hold of, but George persisted. He offered his help with any patent law matters related to his research. Tom was hesitant at first because most of their patents went through the university's legal office. However, Tom did not seem to be very excited about the support the university was providing.

"I am not sure what you can do for us. First, we are not ready yet. Secondly, I believe the university's legal department will have to do their thing. However, my colleague Sam might need some help regarding his research."

"Let me just explain the process to you—just as I did in the workshop, but more specifically to gene research. It will be great if Dr. Sam Roy joins us. I think you told me that he is originally from India. India has patent laws, but their laws do not cover biomedical issues in detail as it is done in the US," George tried to convince Tom.

It took several phone conversations, but he has managed to get Tom's consent. A significant credit went to Alicia and Gina, Alicia's cousin. Tom convinced his colleague Sam. Before agreeing to have lunch, Tom insisted that the lunch discussion had to be purely confidential among the three of them.

"You need to meet him. He is the brain behind the new process the university team is working on to fight cancer," Tom told him over the phone. They showed up thirty minutes late, but he had no difficulty recognizing the pair.

"Dr. Zazac?" George extended his hand to Tom. Before he could say anything, Sam offered his hand as well.

"I am very sorry for being late. The experiment ran longer than expected, and none of my students were around to take over. I could not leave it unattended. I hope you understand." Sam's intelligent eyes were fixed on George's face. He was dressed in modest clothes and a heavy parka. His hand felt gentle in George's rough hand. George realized an immediate bond with Sam. Perhaps, it was the sense of being different or perhaps he knew that he would be the right person to help Sam to protect his intellectual property.

The lunch went very well. Like George, Sam was fond of spicy food. They kept sprinkling the Veracruz habanera sauce on everything.

"We are on the verge of making a significant invention about curing cancer. It is under testing now."

Sam's cell phone rang. As he responded, George noticed a picture of a pretty smiling woman on his screen. A shy glow appeared on Sam's face and his voice took on a much softer tone. It was not hard for George to understand who the person on the other end was. Ah! Those days of chases—Alicia could not do anything wrong then.

"We need to teach Sam how to chase girls and the dating tricks of this country," Tom commented with a smile as Sam turned off the phone.

"What testing? You mean human testing?" George interjected.

"Perhaps—if everything goes well! We found no adverse effects in mice," Sam blurted out. Tom was not as forthright as Sam. It was clear to George that Sam was on to something big. He was prepared for Tom's hesitance. He took out the two-party non-disclosure agreement he prepared for this occasion.

"Sam, I do understand your hesitation about telling me anything. Here are two copies of a non-disclosure agreement I prepared. Take them with you, and if you agree we can sign them. Then, you can let me know more."

Tom's face relaxed.

"I don't know too much about patenting. I am only interested in curing people of cancer. If our work succeeds, it can wipe out cancer from the world, just like deadly diseases like smallpox and leprosy have been wiped out," Sam said.

"True. However, the world is no longer simple. Modern times are very different from the time of the good doctor Armauer Hansen, the discoverer of the leprosy bacillus. He died in 1912. Without a patent, your discovery will be copied instantly by somebody else," George replied.

"But, I am not completely ready yet," Sam protested.

"I understand your hesitation. You want to make sure that your process methodology is fool-proof and you need to do so before you file. However, please consider keeping me involved on a weekly basis, if you wish to go forward with me."

"Why such a hurry?"

"I do not know your process, but I am sure to bring this treatment to the market you will need a lot of money. You will need equity partners and venture capitalists. Without a viable patent you will not be able to attract folks with money. Once you get the patent, you can license your patents to others. The patent law used to be somewhat different before. However, since 2013, the US has become a first-to-file patent granting country. Simply said, the person filing a patent as a conceptual inventor, is granted the primary patent right."

Tom nodded in agreement, but George explained the concept further to Sam using more examples.

"Sam, you should be thinking about filing patents as soon as possible after you achieve a measurable success in your clinical trials."

"You should meet again in Sam's labs when he begins his clinical trials on humans," Tom suggested.

"Please do not wait too long. Trust me—since it is a patent related to humans, any lawyer will need to do a lot of preparation," George advised before they departed

Returning to his office, George looked at the framed document on his wall. He had read it before, but now he read it again. It was a copy of the first Patent Act the US Congress passed on April 10, 1790, titled "An Act to promote the progress of useful Arts." The first US patent was on the method of producing Potash. It had many uses. A big use was for making gunpowder. Who knew how many patents had been granted for weapons? There were many patents on medicines as well. Patenting *in vivo* gene treatment with a foreign gene on humans had never been done before. He contemplated the many ramifications of this new work. He needed to let his boss know about it. He asked his secretary for an appointment. He was lucky. His busy boss was intrigued enough and agreed to see him immediately.

"If we get this case at all, it will be a completely new area for our firm. We have to be diligent and careful. Advocacy groups in the country will be against us. Are you ready for it?"

"I have considered it. I believe Sam is not really into making money. I am sure he will agree to make provisions allowing humanitarian concerns.'

"Let us hope so," George's boss replied.

After several months, the long expected call from Sam came. George was elated. It was time for show and tell. George almost got lost in the cavernous halls of the gigantic old building where Sam's office was located.

"Hi, George! It's good to see you again. I have heard a lot about you since the last time we met. Your reputation precedes you." Sam looked elated.

"Look who's talking! I saw your interview on the TV. You are everywhere, man!" George replied.

"I want to talk to you about a very important matter. I am planning to talk to a well-known geneticist, Brian Warten, about a gene therapy trial. I want you to represent me when we file for a patent," Sam said.

"You are doing the clinical trials yourself?"

"Of course. We have been working on a bacterial protein called rhabdosin in our lab. We and others have shown that it can shrink tumors in mice. Another group finished toxicity trials on a piece of another protein. It has 28 amino acids. Not only is there no toxicity, but some tumors disappeared altogether. We will try to start a Phase 1 trial with rhabdosin, as soon as the investors come up with the money and I get the regulatory approval."

"Sounds like a good plan. How can I help you?"

"I am planning something much bolder and bigger than just injecting rhabdosin in patients. I want a more sustainable solution. I have a patient, a young woman, who has mutations in two breast cancer susceptibility genes called BRCA1 and BRCA2. She has developed ovarian cancer. I hope to start treating her with rhabdosin. Later, I would like to insert the gene for this bacterial protein rhabdosin in her bone marrow stem cells that will make the proteins in her blood, in such a way that she will make rhabdosin herself. The rhabdosin made by the stem cells will find the cancer cells, whenever they appear in her body. There is another woman, a friend of the cousin of the first woman, who has the same susceptible genes. However, she has not developed cancer yet."

"Why can't the second woman be part of your future clinical trials?" George said.

"She does not have cancer yet. The FDA won't approve of injecting her with rhabdosin or anything else. Myriad Genetics has tested the first person. They filed for the patent on BRCA1 and BRCA2 genes and mutations in 1995.

"Yes, I know all about Myriad Genetics patents. The Patent and Trademark Office was sued by many physicians, patients and the American Civil Liberties Union in 2009. In 2010, seven patents on BRCA1 and BRCA2 were revoked by a judge of the Federal District Court of New York, arguing that they were products of nature. Now the case has just been reviewed by the Supreme Court. The Supreme Court ruled that the isolated and purified BRCA1 and BRCA2 genes are not patentable but their complementary DNA could be.

"That is very traumatic. Hundreds of human genes and sequences have been patented since the *Diamond v. Chakrabarty* decision in 1980 that life forms were patentable as long as it's man-made or created through human intervention. Lots of treatment modalities will now be on hold."

"Can't you insert the good BRCA1 and BRCA2 genes in her DNA?" George asked.

"That would be much harder. These genes make repair proteins. We don't know exactly how they work. We do know how rhabdosin works. Rhabdosin will only enter the cancer cells and won't touch the normal cells. We might be able to make some common blood protein like albumin with rhabdosin attached to it. I am even thinking of germ-line gene therapy. That way, not only will she be cancer-free, but her children will be able to make rhabdosin and fight cancer," Sam said.

"Wow! Has that ever been done in human beings? I know that Monsanto is doing similar things with agricultural products."

"Not in humans. But the technology is there. They can take the eggs, suck out the bad DNA from them and put in new DNA. Some mice have been altered this way. They can also take an adult person's cells, and change them into cells that can become more specialized cells. If we could put the rhabdosin gene in certain stem cells, called mesenchymal stem cells, they will deliver rhabdosin that can go directly to the cancer cells, like a special courier service. This is where I need help from you. Many people will resist this change. They will sue us. We need the best representation possible."

"This seems like a very promising way to treat people. I will certainly do some research and find out if I can help you. There is now a precedent for this type of gene therapy that lasts for a lifetime in plants and Monsanto has patents for this technology."

"Wow," said Sam. "Tell me about it."

"We have a case of Monsanto making a Soybean plant called Roundup Ready. It has an enzyme called glyphosate tolerant 5-enolpyruvylshikimate synthetase, also known as GP 5 EPSPS," George looked at his notes before naming the enzyme. "The presence of this bacterial enzyme makes the soybean plants resistant to the herbicide Roundup. A farmer in Indiana, Mr. Bowman, used the patented seeds for growing soybeans. When Monsanto became aware of Mr. Bowman's use of the patented seeds to grow the improved soybean plants without paying them any royalty, Monsanto sued Mr. Bowman in 2007. Just recently, in May 2013, the Supreme Court ruled in favor of Monsanto, finding infringement of the patent held by Monsanto. In Europe, there is resistance to genetically modified organisms. In America, we are used to them," George stopped.

"George, I am curious about the Supreme Court decision on the validity of Monsanto's patent on genetically improved soybean seeds that can be grown for many generations. So, if we are lucky to succeed in the germ line transfer of the rhabdosin gene to the future generation of kids susceptible to breast and ovarian cancer, can we patent such a germ line therapy?"

"What type of gene therapy has been successful in humans?" George asked.

"The first case of a cure in United States was a 4-year old girl who suffered from an immune disease called SCID, severe combined immune disease. She was provided with the missing enzyme, adenosine deaminase, using a virus to transfer the gene. Since then hundreds of trials have been carried out. There was a setback, when Jesse Gelsinger, a volunteer in a gene therapy trial died in 1999. But there are several success stories in other diseases. No germ-line gene therapy has been tried in humans yet. We could be the first to try it. That's what Brian and I are going to discuss," said Sam.

"I will certainly study the pertinent cases and give you an answer. There is a lot of potential for treating patients, especially your friend or her friend, using germ line or somatic cell gene therapy," George said.

The meeting ended there, but the trials and tribulations were just beginning.

Spring had began two weeks prior, yet the sky had been overcast for ten days in a row. In spite of the dreary day, George began his lunchtime run. Afterwards, he would get a burrito from one of the hole-in-the-wall stores, buried in the basement of the office building around the corner. Buds of daffodils were coming out of the raised containers along the streets.

His boss seemed quite pleased with him. His initiatives might be paying off. His law firm had received some queries from biotech giants. Other day, at one of the bar association gatherings, somebody asked him about giving a talk on biotech. It seemed that they know something that he did not know yet. It was amazing how news travels fast. He was sure that nothing came out of his office. He did not want advocacy folks knocking on his door yet.

It was a day of reckoning for him. He was torn between his religious faith and his urge to help the world in the only way he knew. Thanks to Sam, he was convinced that there is a possible cure for cancer. That meant that he had a moral obligation to help the process of taking the cure from the research lab to the market. Sam and his institution already received many queries. George wondered if he was doing the right thing.

George could not shake the feeling that if this patent was

granted, he would be at the center of many controversies which generally bring nationwide recognition. Sometimes good, sometimes bad. But as they say all press is good press. Nonetheless, he did take into account the position of his religious faith. His faith seemed to be of the opinion that if genetic engineering process removed human suffering, then it was morally acceptable. However, it was only one interpretation.

His thought turned to his father who spent years treating people for free.

"Follow your heart and do good. Everything will turn out fine." That was his steady advice. George wished that his father could advise him now. However, with dementia setting in slowly, he lost his best friend. Perhaps someday soon, Sam and his colleagues would find a cure for dementia as well.

It was time to return to work and deal with mundane patents that required no particular moral concerns. As he ran back to work, he knew that it was important to save thousands of men and women who fall victim to cancer. Even in the patenting process, provisions could be made to keep a moral balance.

Tonight, he could go home on time.

George let himself in through the garage door and noticed Alicia watching something on TV. Whenever she was in good mood, she came forward and kissed him. In the past couple of months, George was less stressed about money with the prospect of a promotion and so was Alicia. But she did not make any move today, just turned her head and glared at him.

"OK. What did I do wrong again?" George was puzzled.

"Tell me what you did not do wrong. Do you have to sell your soul for money?" She fumed. What a contradiction!

Facing the TV, he noticed the fiercely beautiful face of Uma Thurman floating on the screen and he understood. He saw Alicia was watching the cult-classic Gattaca. He remembered what Jerome said in the movie, "There is no gene for fate." He did not have a gene for fate, at least with his life with Alicia.

"But how did you know what I was doing?" George learned early in his marriage that any detailed technical discussion bored Alicia to death. He was confident that Alicia would not know what he was involved in. In a way, he never wanted her to know about the patent before it was awarded.

"Well, I met your boss's wife at the Friends of the City Gardens this afternoon. She filled me in with your new activities with great pride. I dug up the movie from our stash. You should probably see it again."

thirteen
Sam's twilight party

Sam came home early from work on Friday. He had too many things on his mind. The distraught face of Erin haunted him. "Does she not deserve a second chance, Sam? She is almost as young as I," came back in between all his experiments, conference calls, his lunch time, his shower and the routine work that he had to do … almost like a kind of Halley's Comet returning every 300 seconds or so. Was she hiding something? The self-made woman with all her confidence, pride and inner strength, maybe she does not want to divulge the fact that it is indeed she herself that is the case? If that is true, he has a lot more weight on his shoulders, Sam rationalized.

Then there was all the media attention and the fabulous success that he had attained through his experiments with rhabdosin on mice. It was a matter of months until clinical trials would start on humans. His mission had been faced with a sinister sense of urgency, just due to Erin. He had started taking it personally to be able to save family members of Erin and perhaps even Erin someday. Was he in love?

When he came home, he found a pamphlet outside his door. He had found similar ones at least three times within the last week. Was this just a prank? He had to crack the mystery. What the hell could be the meaning of all those pamphlets?

Summer Jive
"Under summer twilight,
Come one, come all,
To witness long lost sight
To relive a forgotten ball
With friends long lost ..."
Remember the day nineteenth
A mirthful evening earned by stealth ...
(Come to Meijer's on I-90E to look at the joker piebald!)

When the long hand and short hand are totally even
Slightly before eight, and after seven!

Long lost sight, friends long lost, what the heck? Which long lost friend would not contact him through the Internet? Who has the patience to come to his doorsteps three times within a week? And this was no American salesman. The person knows that this riddle cannot be solved by a person of average intelligence. No average American would even waste 10 seconds on a puzzle. Video games on an iPhone are easier ways to kill time. The puzzle was easy to decipher ... roughly after 7:30pm, when the shorthand and long hand of a watch are perfectly aligned. But why this sense of encryption? Was this a trap?

Thus reflecting, he set off in his Mercedes towards Meijer's at 7:15. He had to crack this little puzzle or whatever it held underneath. A scientist must know the root cause. He reached Meijers at 7:30pm sharp. After a quick scan near the entrance, his gaze was caught by someone near the entrance. There was a short man of Indian origin, or maybe Bangladeshi, who was dressed in a very conspicuously multicolored shirt much bigger than his size, and pajamas which were also oddly multicolored and much bigger than the man's requisite size. He approached Sam, almost as though he was waiting for him.

"You Sham?" He spoke with a Bengali intonation on the "*Sh.*"

"Yes, they call me Sam. And you?"

"I am Jeet, the coordinator of the evening party"

"What party? And what is this all about? Why all these cryptic pamphlets? Do you have any idea what kind of important work I had to leave and rush here?

"Sham, Sham ... we all know you are too busy. We are honored that you spared time for us. Please follow me—we have very little time. I promise, you won't regret. Please get in the car," and they were already near a very run down discolored, disfigured Versa that almost looked like scrap to Sam.

"I am not going to any place, unless I know what all this nonsense is about."

The man looked at Sam almost on the verge of tears and he muttered, "Even when you come to know that Anil-da is suffering from advanced stage of colon cancer, and that he is waiting for you?"

Sam was shell shocked! Anil-da! In Bengali, -*da* is added to the name of a respected person. It means elder brother. Anil-da was really Anil. Hearing Anil–da's name brought a barrage of fond memories to Sam's mind. In fact, the only person other than his dad and his sister that Sam admired very dearly from his Kolkata days was Anil-da. His mentor, friend, and philosopher, who always motivated him with books on classics, science, "Physics for entertainment" and "Physics for fun." He is the same person, who introduced Sam to world cinema, classical music, poetry and dramatics. A true renaissance man, who shaped Sam considerably. He used to call Sam almost every day in the early morning to walk for several kilometers together and discuss cinema, poetry, literature and other topics.

Sam was frenzied ... he found himself panicking ... and talking nonsense. "Anil-da, here ... what the hell ... and I don't know yet ... Anil-da has cancer? I didn't know ... and my research is on ... oh God! How come?"

Jeet held Sam's hand empathetically as though holding a kid and ushered him to his car.

"Why not come in my car? I can drop you here when we return," Sam asked.

"That's not allowed. Besides, for the localities we will cross a Mercedes is too visible. This is a secret party." The man drove the car as though he was just learning, with rough braking and acceleration, and with sharp turns. They went through some bylanes of rundown localities near the city's new airport, localities where lots of homeless people set up their mobile homes and shanties. And finally the man parked the car abruptly in front of some bushes in front of what looked like a forest reserve. It was not a parking lot, but Sam did not care.

He almost wanted to sprint. Jeet took small strides and was casually ambling through dusty narrow lanes created by frequent walking on grassy landscape. Sam knew now that it was Anil-da's English in the few lines of the puzzling invitation. His mentor had to keep a touch of class after all in everything he did.

Then they reached a place that was kind of cleared and it was already dark. He could see some silhouettes of three persons. And as soon as he reached, there was a music that played from a tape recorder. A very well known song from his childhood, sung by the legendary singer John Lennon, "Yesterday, all my troubles seemed so far away …"

Then the battery lanterns suddenly went on. And a frail figure came trotting towards Sam. Yes, it was Anil-da. The last time he had seen Anil-da was in Kolkata five years ago. Anil-da would be in his mid-fifties now. Anil-da's eyes twinkled and he said, "Happy birthday, dear boy, Our dear Amar!" and gave Sam a fond hug, even though it was a weak hug. He was not that boisterous. There were two other people who came to Sam and wished him happy birthday.

Sam hardly remembered his own birthday. It was his dad who used to wish him "happy birthday" over the phone every year. But that stopped in 2009 when his father suddenly expired of cardiac failure. His sister still kept the tradition alive, but Ria was travelling in Europe, and was often forgetful. Anil-da had also wished him well every year since his college days.

And lo and behold, there was Nitai and also Subir. Nitai was the fish seller in Kolkata who was so dear to Sam since his childhood. Nitai always gave an extra piece of fish to Sam's dad as lagniappe for Sam, something that usually does not happen here in the US. His dad, an altruistic person had done a lot for Nitai and people like him in the lower rungs of society, struggling to eke out a living. Subir was the snacks seller in Kolkata that Sam has known since childhood.

They had planned a grand feast for him. So that was all what this is about! Sam's eyes welled up with tears. It felt like home again, the warmth of middle class roots, with people and chapters straight out of his childhood, adolescence, and adulthood. These were the people who burnt midnight oil when there was a medical crisis in his family. When his mother expired, these people loaned their shoulders to his dad for months. They came with everything that they could offer. Dinner for his family, when their cook was not there. A quick rickshaw ride when taxi was not available, and attending to the children when their parents had to go and run errands for emergencies in the families of his mom or dad.

Sam asked "Why did you not come to my home? There are many rooms there. You all can stay. And I can treat Anil-da. I know I can."

"My days are numbered, dear," Anil-da replied with his smile that had layers of sorrow beneath it. "I just wanted to see you once before the D-day, and who else are better companions than these two smart guys?"

"Do you have any clue what we have discovered in the field of cancer, and what we are going to do in the coming months? Why did you not tell me earlier, Anil-da?"

Anil-da went on, a bit weakly, pausing for breath every now

and then "We did not want to burden you ... also we didn't know whether you had a girlfriend or whether you were living with someone. Besides, our dress and etiquette may irk your guests. Although I can speak English, Nitai and Subir can hardly even speak Bengali properly, let alone English. They would be embarrassed to be in your posh setting. Boy, there will be world class guys coming to visit you and how can we spoil the show?" To this, Nitai and Subir laughed out loud. As though this difference between their ignorance and Sam's intelligence is something that they are genuinely proud of.

Anil-da was smiling. "We are staying at a grocer's place, who was also a friend during childhood and college. He came to the US during early '80s. Nitai does the fish cutting and cleaning for his shop. Subir does the cooking and helps him in food catering orders. He also works part-time at the donut shop near your home, just to earn some extra bucks and also treat you to the lovely Masala chai, that you adore ... but you hardly went more than thrice to buy that tea. I do almost nothing. Except for reading books and chatting with my friend. We are all like shadows to you; always loving to protect your family ... We owe so much to your dad and mom. My children were practically raised in your house."

"Oh, so that was where the exquisite tea was coming from," Sam mused.

"What about my daughter's marriage?" Subir came with a plate of freshly fried samosas and onion pakoras with a chai. "Your father practically funded it. And not only that, he himself chose the groom ... such a wise man!" Subir bowed his head and folded his hands as though praying for his dad's soul.

Such love, such warmth in this rustic setting. Was this a dream? Sam pinched himself hard. No it wasn't a dream. Nitai had almost made a small five by five feet fish stall. He was cleaning fish and frying them, making curries of fish just like a street side vendor would do in the rundown localities of Kolkata. Subir was helping him too. There were curries, vegetable dishes and sweet dishes, all prepared by them ... all of Sam's favorite dishes. These were his extended families. Not educated, unpolished, but with unflinching love and loyalty towards his family. Giving back with whatever they could, in a humble way.

There was a huge United Airlines flight that flew over, hardly a few hundred feet above them prior to landing. Amid the twilight sky with another flight over his head, it was a surreal setting, his rustic setting amid the setting of the most developed and most powerful country on this planet. It was a little private chink that he didn't want to let go. This was his breather. This was his little oasis amid the entire desert like drudgery and stress in everyday life. Sam decided that he will leave no stones unturned for these three souls, who have always gone overboard to see him happy. They were family indeed, not extended, he rationalized.

He talked to Anil-da, Subir and Nitai, listening to his favorite songs, seated on a plastic cloth laid over the summer grass. Anil-da also explained that he had been sending Subir in a proper dress in front of Sam's house every day just to make sure that he was fine. He had also sent Jeet several times to mail the cryptic pamphlets to arouse Sam's curiosity. Anil-da knew only too well that anything that is like a puzzle would definitely get Sam's attention. And the whole idea was to give Sam a surprise birthday party, a party that he will remember forever. Anil da also explained that he had spoken to the owner of the local donut shop to try and add a special masala chai so that all the neighboring Indian, Bangladeshi and Pakistani diaspora would love it. It is something novel and the owner welcomed the idea. No wonder, the masala chai was something not only to titillate Sam's taste buds but also to give a component of his long lost past at hand's reach. Sam listened in awe. Such was the meticulous planning of Anil-da.

"From tomorrow, all of us going to stay at my place and I will treat Anil-da myself."

Anil looked at him perplexed. "That cannot happen."

"You have a tourist visa. I can always treat you. Jeet told me

that you are in stage III. I believe our potential drug rhabdosin
will work well on stage III or IV patients. I am pretty certain I can
crack this. Will you be willing to let me try my research on you?"

"Son, I have come to die peacefully. I have led a good life. No
regrets whatsoever. But if you can benefit from this research on
me, I am game … anything for you, boy!"

Sam went and hugged Anil-da. Sam was crying like a child. He
did not quite understand whether it was tears of agony or ecstasy.
But he felt at home. Very much at home.

fourteen

Vacation

After the dinner in the Indian restaurant, when she had impulsively confided in Sam about her friend's mutations, which were really her own, Erin was sorely assailed by pangs of conscience. Should she have just told him that she was not what she appeared to be, a healthy normal young woman, but a disease time bomb, waiting to fall into a fatal illness at any time? No, she couldn't bring herself to admit that, for her feelings for Sam had made her yearn for a new lease of life. His lustrous eyes when he talked about his discoveries and his plans for the future made her believe it was possible to hope for a disease free existence. But she couldn't stop blaming herself for not having the strength to tell him the truth about her own state of health. All she knew was that she had lied to him for a reason. And that reason was that she was crazy enough to fall in love with this crazy scientist and his fierce optimism in the face of serious odds. That is why she had told him of the mutations to find out if he thought her friend, that is herself, had any hope of survival. She did come away lighthearted that she had a chance after all but soon realized she couldn't proceed any further without divulging that it was she herself who had the mutations. She would have to figure out what her next step should be. The truth was that Sam was not her doctor, just an independent researcher. How long could she bother him in third person by talking about her friend? He was just too smart. It was a matter of time before he cracked her secret. Or worse, she could let slip something that would make him find out. It was much better to just stay away and not answer his texts.

So it was by chance that in early fall Erin ran into Sam at a local auction of rare research books. In dark jeans, an open neck green polo shirt and a dark blazer with a bright striped scarf wound playfully round his neck, Sam looked cool and chic and on a mission. "Hi Erin" he said evenly when he saw her, then asked with a twinkle in his eye, "You look great! Been exercising a lot lately? You seem to have lost a lot of weight. Is that why you didn't get a chance to call?"

One part of Erin knew it was just an observation but Erin was on edge right away and panicked, that means I have become too thin, like Dr. Lee said I might. Her mind kept saying over and over, he thinks I look sick, sick, sick.

"Hi yourself," she replied, trying to look nonchalant, and then turned away rather stiffly to pick up a few treasures. She was unable to make any normal conversation with Sam without hypothesizing about her body. Suddenly Sam said, "Hey, don't look so serious! See it's such a nice day?" Seeing that she was engrossed in her book, he continued browsing around some old medical journals and said casually, "Remember when we were talking last time, I told you I was thinking about visiting my relatives in India. Didn't you say you could do with a vacation yourself?" We? Talking about vacations? Where and when? As Erin was puzzling out a sharp reply, Sam continued, "Don't you get my texts? I thought maybe you got a different number or something. I was about to email you one of these days, about this vacation deal my friend sent me. See, every winter, my friend emails me these awesome vacation packages. He has a timeshare in Maui, Hawaii. He keeps asking me to go visit. But I've told him, only if someone shares the expenses, then maybe I could give it a thought," he ended seriously nodding his head.

Erin couldn't help bursting out in a giggle. She put down the heavy volume she was trying to read and turned to face Sam with genuine mirth, "Are you kidding? You really want me to believe you need someone to share your vacation cost, Sam? A well established doctor and research scholar like you?"

Sam appeared delighted and answered instantly, "See, I got you to laugh at least, didn't I? And Erin trust me, you look much more beautiful when you're happy. Worry lines don't suit you at all, my dear, Bonita!"

Erin made an effort to draw her defenses back up again. For some reason Sam's flirting and teasing meant so much more to her than just that.

"Anyway, you got me again, smart lady! So let me come straight to the point," he continued easily, sobering up. "Would you like to take a vacation in Hawaii? It'll be quite a good deal if we split the cost of my friend's timeshare and vacation package. Are you game?"

Hawaii! Erin remembered the colorful brochures she and her mom used to go through as they waited in the doctors' chambers so many years ago, planning their dream vacation after her mom got better. It had never happened for Nancy but wouldn't she be very happy wherever she was, if her daughter fulfilled her dream? Erin had the feeling that things were happening too fast but for once she didn't mind. She felt tired of being afraid anymore and wanted to be free of all ill thoughts. She looked at Sam. There was an open challenge in his expressive eyes. Was he really asking her out on an exotic vacation date? But she didn't want to think through it all now. She threw all caution to the winds, and meeting his eyes steadily replied quietly, "First I have to see when I can get some time off."

"I'm calling my friend right now" said Sam quite thrilled and reached for his cell phone matter-of-factly. As he spoke in a language unknown to Erin, she watched his features, as surprise, laughter and mock anger flitting across them. Behind his dark head, she could see the riot of gold, brown and auburn of the late autumn leaves outside. It was unlike any experience she had had before and she enjoyed it so much she forgot about herself for a while. That was when Erin knew for certain she would have to go on this vacation with Sam. She still had misgivings about where their budding friendship would lead, for it seemed to have tremors of being more than just that. For however much she may have been in denial, Sam had started to mean much more to Erin than even she could fathom. So she couldn't give up the temptation of seeing it to the end.

They decided on the shortest and least expensive trip: three nights and two days. They took the same nonstop flight to Honolulu but since both had requested window seats, they sat in different areas. Which was ok, for when Erin glanced over occasionally, Sam was always listening to lectures, as he said, with his ear phones plugged in.

They took a short connecting flight to Maui and then rented a car to drive to the timeshare. They had planned to share the rented car also and so Erin asked "Do you want to drive?"

"Since we've planned to share almost everything on this trip let me drive us to our condo and you can drive us back to the airport when we leave."

"Wow!" answered Erin "Are you always so calculating?" as she climbed into the passenger seat after putting her backpack in the trunk.

"I grew up in a tough neighborhood in India, Erin. I had to learn to survive. Wouldn't you like to know some tricks too, about surviving, I mean?" He turned his full attention on Erin and in the tropical light, so close to Sam, she now saw that his eyes were not dark at all but a liquid brown with flecks of dark gold deep down. She blinked and looked away quickly, but couldn't get rid of the feeling that he was watching her closely and hinting at something more meaningful.

Sam had given her the map, for he said he was not good at reading them, and between the two of them they arrived at their timeshare in the late afternoon. It was by a rather secluded beach. She couldn't take her eyes away from the turquoise green Pacific waters as they lapped lazily on the rocks close by, sending a multicolored spray into the sky. The sun was getting ready to set and the palm trees caught its message of colors and laughingly waved their rustling response.

"I know it is beautiful" said Sam looking at her with laughter in his eyes, "but don't you want to see your room?"

She had made it clear from the outset that they'd have separate rooms, and she nodded turning to look for an elevator for she remembered their suite was on the eleventh floor.

The first thing Erin saw when Sam unlocked the door, was sky and water and the sun as it dipped into a blazing liquid of unimaginable shades of Prussian purple, cayenne pepper and baby pink. She dropped her backpack and ran towards the open balcony. Suddenly Sam was in front of her and she almost fell backward as she stopped in her tracks.

"I thought we Indians don't know anything, can't see anything and smell of curry. But it seems you too have a problem and can't see a big thick glass balcony door right in front of you. The last thing we need is an accident, right?" Joked Sam, as he moved forward quickly and opened the balcony door for Erin.

Erin laughed out loud and as she did the sound of the waves on the rocky shore rose to her ears. She turned to Sam and said with genuine gratitude, "Thanks for bringing up this idea of a Hawaii vacation. My mom and I had always wanted to come here. "

Her eyes grew misty and Sam couldn't tell by looking at them which was a truer blue, the sky, the water or Erin's eyes. When she turned back inside, she was sure she could read a deep vibration in Sam's eyes. From a young age Erin had learned to watch out for desire in men's eyes but she was sure this wasn't the same. Sam ended the moment by declaring practically, "I don't know about you but I'm starving. So let's go eat."

"Everything looks so expensive here," replied Erin.

"Ok, let me take you out on the first night …"

"And," Erin finished, "I can take you out on the last. But we'll have to pick up something from grocery store downstairs for the one night in between." Then as another thought came to her mind she continued uncertainly, "If we are going out for dinner, should I change?"

"Oh I'm very informal, as you may have noticed. Just come as you are. You are naturally beautiful," said Sam and then quickly looked away. She smiled. His sudden shy demeanor unconsciously endeared him to Erin all the more.

"Ok then," she said as lightly as she could. "Let's go."

At a rather nice beach restaurant, Erin, much to Sam's amusement, ordered the cheapest fish taco on the menu but Sam insisted on an expensive white wine. As they ate and drank in the open, they could see that the sun had disappeared into the water leaving strands of color floating on the waves. It was as if a light bulb had been switched on under the waves making the whole ocean luminescent. Then the gold and the vibrant candy colors were replaced by warm maroon and navy tones. Soon a bright half moon rose and bathed the sky and waters with sparkling silver sequins. Erin's heart ached, thinking of how she would have loved to show her mom such a beautiful world. As it became dark inside and the lanterns were lit, the flickering flames etched shadows on Sam's face and he touched lightly Erin's hand on the table, "I too think of my mother a lot. I was her only son and she had high hopes for me. But I know my mother's love and blessings are always with me wherever she maybe."

Erin felt oddly consoled by his thoughtful remark. But she couldn't get rid of an instinctive hunch that there was a reason he had brought her here. It wasn't just an ordinary vacation. Would she ask him outright what it was, and ruin the lovely evening and whatever it was that was building up between them? She hesitated. No, she could sit there for the rest of her life watching the light play on this genius of a man who would hopefully save her cousin

Kathy from certain death and would start his scientific play on her too one day, if he only knew.

Their respective bedrooms were on two sides of the center sitting area of the condo, and as they bid each other goodnight, Sam said, "It's a beautiful world and …" he stopped and looking meaningfully at her, continued with emphasis, "all beautiful things have a right to enjoy it for a very long time."

Had Sam guessed that it was she who had the mutations and not her friend as she had led him to believe? Then why did he bring up this subject of coming on vacation? He seemed too genuine to have any desires for Erin. The realization made her ponder rather indignantly, just what he thought of her. Could Sam Roy get the smart and beautiful Erin Goldberg so easily? But when Erin fell asleep, listening to the waves, there was another realization that was creeping into her mind, resonating with the wind and water. She knew it was not that hard for a very Western girl to fall in love with an amazing scientist born in a oriental land.

The next day and night passed in the whirl, as they enjoyed visiting the volcano parks and lazed on the beach. But on Saturday Sam seemed really tense and told her to go shopping on her own, if she so wished. He didn't come out of his room for breakfast and when he finally stepped out late afternoon, when Erin got back after picking up a few souvenirs for her aunt and her cousin, he seemed restless. He looked unkempt and started pacing up and down the open sitting area. To lighten the atmosphere Erin said, "Remember it's my turn to take you out for dinner tonight."

He answered tersely, "I'm not hungry, I just had some leftovers from last night's dinner that were in the fridge."

Erin caught his mood and so as not to add to his distraction she too replied lightly, "I had a rather big fish salad too! Maybe I'll go for a walk later!"

Without really answering her Sam quickly took a few steps towards the open balcony and continued, "I can't keep beating around the bush anymore. Erin, I asked you to come here so I could propose to you under the Hawaii sun and stars!"

Erin had to repeat, "What? What do you mean?"

Sam turned and looked earnestly at her and began, "When you told me about your friend having those mutations, didn't I tell you she could get cured from the medical threat with my experimental rhabdosin?"

Erin was startled. She had been suspecting this all along. So was he trying to get to her friend through her, by being nice to her? Was he just trying to wheedle her into making her friend the subject for his subsequent experiments? But she didn't think Sam would turn out to be such a selfish, ambitious, intellectual pig, trying to get his own job done in whatever devious way possible! Nor did she think that their feelings for each other would be upturned so soon!

Erin began decisively, "Dr. Roy, because I told you about my friend's condition and because you think you can help her, it doesn't mean she will agree to your experimental drugs. Besides I'll tell her not to, seeing how crooked your motives are."

Sam cut her off, "What are you talking about. Erin? What does my proposing to you have anything to do with your friend?'

"You tell me," breathed Erin quite beside herself with wrath at this feigned innocence. "Proposing to me? And then talking about my friend! Why? It is not happening, I tell you. You are not going to make my friend your guinea pig. She is not yet sick. Yes, if she has cancer, maybe she'll come looking for you, Harry Potter of modern medicine, and you can claim to give her your magical potion to try to make it better, but by proposing to me you're not going to make me influence her to come see you! You can be quite sure about that!" He was ruining everything and she was not ready for it to be over between them. She turned helplessly to the fast darkening balcony door and whispered indignantly, "That is final."

Sam moved around her so he could look at her. His face looked harrowed. "Please let me finish. You didn't hear me out. The way you told me the details that evening about the mutations, I knew it couldn't be your friend, it had to be you. I can prevent your disease from ever developing, Erin. Let me help you … why won't you let me? Why do you keep misunderstanding me when it was you who had to lie about some friend? Why can't you trust me and let me help you? Because, I'm different? Why, why?"

Erin came right to the point. "Ok Sam? Now you know! It is me! And I lied that it was my friend because I didn't want your pity." Erin breathed in angrily before continuing, "And now Dr. Sam Roy, now you tell me, why do you really want to help me?" She stuck her finger into her chest so that it hurt. "I'll tell you, why. Because I am the best, most gullible single subject you can find and you can prove your theory through me, by pitying me and scaring me that I'll be sick soon. You don't care what I think or feel, but your job will be done, right?"

Then Erin stopped short. She was panting hard and her head throbbed. Tears started smarting her eyes.

Unexpectedly Sam seemed to give up the fight. He abruptly turned and walked away from her, caught his head in his hands, and said in a broken whisper, "Do I have to spell it out, Erin? In the last couple of days, I thought you would know. You would know how precious you have become to me, over the last year or so." He turned to face her. "The way you've braved all the odds, the storms you've weathered in your life, the way you're still trying to keep your head above the raging waves of things beyond your control. Never complaining, always smart and strong … ready for a fight."

"But I broke down that day in the library when I first heard of my mutations. And I always thought you would pity me for my … my … condition," Erin faltered.

"That just made you more human to me. More … dear." He turned, searched her face and said, "You still don't believe me, do you?" Then he fumbled in his jean pocket and brought out his phone and turning it on the saved mail page, he read in his unique accent and slightly hoarse voice,

For you,
could I be the peace and gold of the sunset
To play within your veins like an infant swimmer and nothing
more?
For you, could I vow to be those flowers reaching out their
hands from the rocks
rewarding you with their tender touch?
For you could I be the skin and color of your iris
A chimerical journey of togetherness come rain, come storm
And nothing more?
Yes, for you,
I could wait for an eternity,
Just so there is a chance of a beginning once again,
After what seems a chasm of nothingness,
And nothing more …
For you, yes … for you …

Sam paused briefly and continued, "I wrote the first part right after I read your "Sunset" poem on the blog, but completed it just this afternoon when you were out shopping. Do you know now Bonita, why I'm proposing? Do you trust me now?" He ended abruptly in a barely audible whisper and just stood there looking at her with so much askance in his eyes, Erin felt herself tremble.

She had never known a more polite, kinder, more lovable man. A more aggravating man too. He always had this terrible tendency of putting the cart before the horse. Why didn't he tell her he loved her right at the start instead of talking about her friend? That had made her leap to conclusions too fast! Should she say sorry? No, that would be too inadequate after such a beautiful poem. She didn't know what to utter, and moved to the wall for support and in so doing brushed against Sam's arm. The electricity of that touch seemed to charge the air between them with sparks. Erin was not sure what happened next. Sam took Erin in his arms and there was nowhere else she would have cared to be. All she remembered was Sam whispering, "Are you okay with this?" And she kissed him in reply.

When Erin awoke it was early morning. They had made love on the floor of the sitting area, to the sound of the wind and moonlit sea with the balcony door open. Sam was still asleep, one arm over his eyes. She kissed him gently on his lips and smiled. When he woke up she knew what to tell him. She put the coffee pot on and went to take a shower. They were leaving that day so she packed her few things and came back to the sitting area.

Sam was up and handed her a cup of coffee the way she liked it, lots of cream and no sugar. Once again she was amazed at how different he looked now, showered, shaved and impeccable. "Erin," he began with a new tenderness in his eyes and hesitancy in his voice. "I don't want you to think what happened between us obligates you in any way to decide, but …"

She caught his hand tight and looked into his eyes, "The answer is …"

Sam quickly moved forward and put his hand on her lips. "Please let me tell you the whole truth first before you commit yourself."

Erin took a step back and looked genuinely puzzled. "What are you keeping from me now?" she asked. "Do you already have a girlfriend or a wife or what?"

Sam made a face. "It's worse! I've applied for the experimental scientist lead in a brand new facility in an island up north. They are looking for scientists. They were ready to let me conduct my experiments with a huge grant because they want their island to become famous for new ground-breaking research by attracting the best brains from the world over. I have not told anyone else. I just found out while you were in the shower that they are ready to hire me as their full time lead scientist and I have to join soon."

"Congratulations, Dr. Roy! One more feather in your cap, or should I say, crown?" Erin laughed happily.

Sam looked serious. "Erin," he began slowly "you're missing the point. This place, Bluefrostland, is a small Arctic island where I may live for many years, maybe the rest of my life, to conduct my experiments."

"Not a bad idea away from the hue and cry of busy cities!" replied Erin clearing up the kitchen utensils. "And why may I humbly ask are you making such a big drama of it, Sam? If you remember we have a plane to catch!"

"This island, Bonita, is in the middle of nowhere, and," he paused briefly, swallowed and continued, "you may end up there too. That is if you commit to being with me, for I think you were going to say 'the answer is yes' just a few minutes ago before I stopped you, right?" Sam ended softly, cautiously.

Erin stopped putting away their coffee cups and stood very still near the cupboard with her back to Sam. It slowly dawned on her that she would probably have to give up her job, her career, her life and start over in a far away land if she said yes to Sam. Sam sensed her change of mood and put his arms around her waist from behind. He was waiting for her next move. Erin stiffened and moved away from his embrace. Without meeting his eyes she said lamely, "It was real … nice … of you to let me know your future plans. Honestly all I can think about now is that if we don't leave in ten minutes we may miss our flight back."

Sam immediately sensed her reaction and moved away quickly, saying matter of factly, "Of course! Let's pack up and we can talk on the drive to the airport."

They quickly paid the bills at the hotel lobby and climbed rather awkwardly into their rented car. Erin couldn't help thinking, nothing in her life could ever be simple and normal and smooth. First she had to fall in love with this weird scientist and now he had this weird notion of moving to some weird island and the weirdest part was she would probably be heartbroken forever to have to bid him goodbye!

On the flight back Sam tried to make conversation, but Erin

was not in the mood for frivolous small talk or work histories. Eventually Sam started looking up some official grant papers and Erin watched the cheesy in-flight movies.

When they parted at the airport parking garage Erin tried her best to call up the magic of their last night in Hawaii but couldn't. She felt cold inside. With usual insight, Sam sensed her withdrawal and gave her a quick farewell hug. "I hope you'll remember the fun time we had in Hawaii. I do think we are great together but you're very much a city girl and so I had to warn you before you took the leap."

"Thank you for your concern, Sam. And I have to admit I really had ... had ... a very memorable vacation ..." Try as she might Erin's voice faltered and trailed away. She failed to see the tension lines throbbing on Sam's temple for she was really at a loss for words battling with her own emotions and common sense. Silent tears cruised down her cheeks but she didn't have the energy or care to brush them away.

Sam looked at her briefly and suggested suddenly, "I have an idea! Why don't you come visit me in Bluefrostland, one weekend in early spring! You can decide then. Just give yourself some time Bonita," he ended brushing his fingers uncertainly on her wet cheeks.

Erin couldn't trust herself anymore. "I'll text you later, and thanks for everything," she mumbled as she quickly got into her car and drove off without even waving goodbye.

Sam had read somewhere that there are times in one's life when joy is balanced with pain. This was probably one of those times for him. He was thrilled about this opportunity to conduct his experiments without a hitch on Bluefrostland. He had this gut feeling that he had finally moved closer to his dreams, daring though they were. His excitement and thankfulness knew no bounds.

On the other hand, he knew all of this would be meaningless without Erin. Not only because he longed to see that she would be free of the threat of a dreaded disease, but also because he had never really known anyone who accepted him so fully just as he was, with all his eccentricities, and made him feel so human and complete. Nevertheless Erin was headstrong and impulsive and though these qualities endeared her to him, he couldn't help worrying if they would also make her break up with him. Was their love for each other strong enough to overcome yet another hurdle? Well, only time would tell!

Sam had confided in Erin, something he himself wasn't sure of, until he had actually blurted out his plans to her. He had applied for a grant for research in a remote island, in response to a challenge. This island had become rich with newly discovered oil. He wanted to do some clinical trials there with the bacterial proteins he had discovered and those he was about to discover. He wanted to do them in collaboration with other scientists. One such scientist was Brian Warten, a famous geneticist. Brian Warten was the keynote speaker in the morning in a meeting in Washington D.C. The day after returning from Hawaii, Sam flew from Chicago to Washington D.C. He took the train directly from Dulles airport and went to Brian's seminar, before he even checked in the hotel. Sam was the speaker himself in the afternoon. Sam was very anxious to learn about Brian Warten's gene therapy trials. He had already made an appointment with Brian a week earlier.

Brian Warten was a remarkable man. Just like Sam he had overcome many adversities before he made a name for himself.

fifteen

Brian's story

It was a blustery September morning aboard a ship approaching New York harbor in 1940. It was a rickety old ship, filled with German Jews escaping Hitler's Germany. Frederich Württemberg, a pale skinny young man with a scraggly beard, was intensely looking at Lady Liberty holding her torch. He was also secretly holding a lamp for another younger lady aboard this ship, Claudia Busch. She had thick braided hair and thicker black-rimmed glasses. He was stealing glances at her every few minutes. They came from neighboring towns in Germany, but they had only met each other on the ship. They chatted often during their slow and long journey. His face was wet. His eyes were shining. He looked like he was crying, maybe because his parents were not with him. He was about to reach America, the land of freedom, where he would no longer have to hide. It should be a happy and exciting moment for him. But he also felt sad. He might not see his passenger friends, particularly Claudia, again after the ship docks.

The immigration line did not move swiftly. A burly man with strong biceps yelled at Frederich. Frederich tried to pronounce his full name, but the man did not have patience. "Warten sie bitte," (please wait!) said Frederich, looking for a shipmate who could speak English. The burly man waved him on and wrote his last name as Warten. Frederich Württemberg would now be known as Frederick Warten. The young lady Claudia Busch moved on smoothly. Only, she had lost something. She was now Claudia Bush. He shyly waited for Claudia. An older but well-dressed man, Claudia's uncle, had come to receive Claudia. Frederick shook her trembling hand and said, "Kiene sorge! Auf wiedersehen. Bis spater" (Don't worry! Goodbye and see you later.). It did not strike him to ask and write down her uncle's address, so he could contact her later. They went their separate ways.

Walter had to wait outside a few more minutes. No one had come to pick him up yet. A few moments later, a middle aged man with salt-and-pepper beard came to receive him with a placard with Frederich's name. They had never met before. Frederick ended up in a farm house of one of their family friends, named Walter Hubert. He was about 50 yrs old. The farm was situated picturesquely on a green-carpeted valley in Denville about 50 miles from New York City. He knew nothing about Walter. Walter had lost his wife a long time ago and had no child of his own. Having Frederick with him, he was overjoyed. Now, he did not have to spend time alone. While driving back to his farm, Walter broke the silence.

"Are you alright?"

"Yes," replied Frederick.

"Do you think you can stay in the US without your family and friends?"

Frederick, in his broken English, responded, "I will try, I have no other choices. Thank you, for taking me in." He paused. "Do you have any children? Family?"

"No," said Walter, "I live alone."

Frederick was in no mood to talk. They reached home, and Frederick grabbed the small suitcase he brought with him, a few clothing items inside, from the car and followed Walter inside the house. Walter showed him a room. "This will be your room," Walter told him. It was a smaller but a lot nicer room than the one they had in Germany. He also could not believe that this would be his own room. He had to share rooms everywhere when they moved around in Germany.

It was around 7 pm. He was tired and hungry as well. Walter told him to take a shower and then come to eat dinner. It was a simple dinner with pasta, salad, baked potato and roasted chicken. Frederick ate dinner, but could not stay awake any longer, and went straight to bed saying "Gute Nacht" to Walter.

Next morning, Walter knocked on the door to wake him up. They had some breakfast—farm-raised hen eggs, bread and coffee. Later, Walter showed him the farm. He enjoyed the tractor ride. It was a small farm of a few acres. A small area was used to raise chicken and the rest was used for corn and different types of seasonal crops, fruits and vegetables which he sold at the nearby farmer's market. Walter asked Frederick if he could help on the farm. Frederick agreed. But, Walter assured him that he needed only a little bit of help and promised him that he was otherwise free to do whatever he wanted do to advance his career.

For a while, not knowing clearly what he wanted to do, Frederick helped Walter on the farm during the day. There was not much to do in the evening except for chatting, playing board games with Walter, or watching TV. Claudia was on his mind as well, and he asked himself, "What was she doing? Where was she living? Did she still remember him?" He couldn't excuse himself that he wasn't brave enough to ask her uncle's address in New York. He remembered that Claudia liked music and she wanted to finish her studies. She wanted to be a teacher.

One day while helping Walter clean the house, Frederick found a couple of guitars stored in the corner of one room. He started coughing while cleaning the dust which had accumulated, perhaps for many years. He cleaned one of them and started playing. He had a melodious voice and he had learned how to play the guitar when he was in Germany. Walter was surprised to hear the beautiful song that Frederick was singing with the guitar.

"You have a beautiful voice. I did not know that you are such a great singer. Where did you learn to play and sing so well, Frederick?" Walter asked him.

"My father," Frederick answered.

Frederick missed his father and there was no contact between them. He was afraid to write a letter to him. Also, he didn't know the latest address either. May be they were still moving from one place to the other. He stopped the music thinking about his family. A question from Walter alerted him.

"Frederick, what are you thinking?" Walter asked.

"Oh, I am sorry. I was thinking about my father, my family," Frederick answered. Walter realized that he was missing them very much. So, to comfort him, Walter picked up the other guitar and started playing. He had stopped playing the guitar since he had lost his wife. It sounded a bit rough.

Frederick realized that Walter knew how to play but had not practiced for a long time.

"I am not good any more. I used to be pretty good, you know," Walter said.

"So, why did you stop playing?" Frederick asked.

Walter paused for a moment. "I did not feel like playing from the day my wife passed away. She was a very good singer as well," Walter answered. Walter lost his wife 15 yrs ago.

"Oh, I am sorry. But, now that I am here, we both can play together," Frederick said.

He encouraged Walter to play with him every day. So, the days passed with work in the farm, and at night they played music. Slowly blustery cold days of winter were passing and spring was knocking at the door. Seemingly dead-looking trees were now live and healthy with fresh green leaves, birds were chirping everywhere.

After several months, Frederick decided that he would go to school. He asked Walter if he would support him to fulfill his dream. "Of course, there is a community college nearby. You can go there," Walter told him. The next day, Walter took him to the school. But he had to pass the English test before he could be a student. With help from a teacher friend of Walter, Frederick became proficient enough to become a student. He wanted to major in music in the college. Walter became his second father and Frederick became the son he never had. They had a great time together.

Claudia had gone to her uncle's home near New York City. Her uncle had a family of four. They had a son and a daughter, 3 and 5 years younger than Claudia. Three cousins became good friends. Her uncle had a small business and he could afford a modest life. He helped Claudia join a university in New York. She wanted to be a mathematics teacher. Her days were spent studying, practicing piano and spending time with her cousins during her spare time. She was extremely good both at school and in playing the piano.

Several years passed. Frederick and Claudia lived in New York state several miles apart, but there was no contact between them. One good evening, as luck would have it, they met again in a concert organized by the German Institute of New York. They had both come to perform there. Claudia saw someone resembling Frederich, the person who had befriended her in the ship, but he looked a bit heavier and he had longer hair. She came forward to say hello and to find out if she was right. She was hesitating, but went forward anyway.

"Are you Frederich? Haven't we met on the ship coming to the United States? What are you doing here?" Claudia asked him in one breath, excited to see him again.

"Yes, I am. I came here to perform in this German music festival. What about you? You haven't changed much," Frederick answered with equal excitement. He was actually quite delirious with joy to see Claudia after so many years.

"I came here to perform as well," Claudia answered.

"Fantastic. I am sorry, I did see a name Claudia, but I did not realize that it was you. I thought you were Claudia Busch?" Frederick asked.

"Oh, that's a long story. But, I don't see your name in the brochure," Claudia said.

"Yes, I am here" pointing to his name, Frederick answered. "But, that's not you. It is Frederick Warten," Claudia wondered. "That's me now," Frederick answered. They explained to each other how their names had changed at the immigration desk and they had a big laugh.

This unexpected meeting became the new dawn of their life. They both finished schools. She had a bachelor's degree in math and he had a degree in music. Claudia found a teaching job at a middle school only few miles away from where Frederick lived. Frederick did not find any job. But, he started his own music school. He started to teach vocal and instrumental music to any one who wanted to learn. After several months, they got married and lived in a small house that both Walter and Claudia's uncle helped them buy. They spent the days at their jobs and the rest of their time they reminisced about their days in Germany, singing songs or playing music together. They both were extremely good students in school in Germany. But, they could not finish school because their parents kept moving from one place to the other to avoid being captured by Nazis. Eventually, following his parents' advice, Frederick started planning his escape from Germany. It took years of planning, deception, and good luck. He made it to the US. Unfortunately, his parents had to stay behind as part of deception. Claudia also came at the same time. She had no family members alive in Germany.

Eventually, they had two boys, Roger and Brian. Roger was two years older. As they were growing up, they became best friends. Roger was always protective of his brother. Brian always tried to follow his older brother. Frederick and Claudia raised them in the Jewish faith, and wanted to give their sons the best education possible. From early childhood, the brothers became interested in music and started to take music lessons from their parents. Brian was very good at playing piano, and Roger was a vocalist. But with constant insistence from their mother they had to study hard as well. They started to show their talents in other subjects as well. The parents couldn't be happier.

From childhood, both Roger and Brian wanted to know about the country their parents came from. So, Claudia and Frederick spent a lot of time telling them the difficult time they had in Germany, why and how they escaped from Germany, why they had to leave their grandparents, and why they could not finish school. The real-life stories of their family made a huge impression on both Roger and Brian. Frederick's father wanted to be a doctor helping people. Claudia's mother wanted to be a scientist. Roger and Brian made up their minds to fulfill the unmet dreams of their parents and grandparents.

Finding time to play music after their studies used to be easier for the brothers. But, starting at age 10 or so Roger started developing problems. He was not doing well in school. He complained that he couldn't see the blackboard or hear the teacher well. Initially, the teachers thought he was faking and that he lost interest in learning. Later, the teachers were concerned and informed the Wartens that they should have him tested for vision and by an ear, nose and throat specialist. The doctor told them he needed to perform other tests to find the problem.

With whatever money they had, Frederick and Claudia had Roger tested. Devastating news came. Roger has developed Adrenoleukodystrophy (ALD)—a rare X-linked disease which comes from the mother's chromosome. ALD is a rare genetic disorder, caused by mutation of a gene called ABCD1. The defective gene causes accumulation of very-long chain fatty acids (or simply, fats linked together many times) and covers surrounding nerve cells in the brain and causes progressive dysfunction of the adrenal gland. The most severely affected tissues are the myelin in the central nervous system. Most common symptoms include seizures, impaired motor function and vision, muscle weakness, mental retardation, emotional instability, restlessness, retinal degeneration and blindness. There was no cure. The disease progresses slowly over many years with fatal consequences.

The family was devastated. A genetic test showed that Brian's DNA was normal and had no mutation in the ABCD1 gene. Brian did not quite realize the impact of the disease immediately. However, as he was observing his brother's very slow but definite deterioration, he began to get depressed. He tried very hard to comfort his brother by playing music together. "Can you sing a song, Roger, while I play the piano with you?" asked Brian. "No, I don't feel like singing," replied Roger. "We haven't played music together for a long time, please, Roger," Brian insisted. "Okay, only one," says Roger. They ended up singing a few more.

Watching the brothers playing music, both Frederick and Claudia felt good. By the time Brian finished his high school, Roger had lost almost 90% of his vision, but he could still hear. He couldn't walk on his own, but he could walk with someone's assistance. He could remember and recite all the poems or sing songs he had learned before the onset of the disease. But he was having difficulty learning new things. They sang and played together whenever Roger agreed to participate. Brian made up his mind that he must go to medical school and do something to help cure this type of debilitating disease. He just couldn't watch his brother's suffering.

sixteen
Brian's professional development

Brian studied intensely to prepare himself for medical school. He also spent hours polishing waltzes, sonatas and nocturnes to keep Roger entertained. At college he wanted to become both a physician and a pianist. After a few years in college, and away from home, he began to think less and less about the family or his brother. He started taking classes in both medicine and music. However, his passion for music became so strong that he earned a degree in only music. He started teaching piano and performing in New York City. But, time demands and performance anxiety made him unsure about his career choice. Then he remembered his days at home, his brother, his parents, and he decided to join medical school again so that he could keep the promise he made to himself.

His reasoning was simple. To be a great doctor you need to be able to not only see patients, but also to develop new therapies for diseases that are untreatable. So you need the mind of a scientist as well, he reasoned. To be a great scientist, on the other hand, you must know how the human body works, how it reacts to infections, and it interacts with its environment. Who knows a human body better than a doctor? But, can you study these two apparently demanding degrees at the same time even if they are thematically linked? He had a hunch that it may be possible. So he went to one of his biology professors and asked if such a possibility exists. The professor sat him down and told him that not only does it exist, but only the brightest and best take advantage of it. "I must tell you that it is highly competitive and demanding, and you must make sure you are committed to this grueling 8 year program. 2 years of medical school, 4 years of research, then finish the last two years of medical school."

"How do I know, if this path is for me?" Brian wondered.

The professor responded, "Why don't you shadow doctors at our medical school and see if you like it?"

"That's a great idea. And I know I will like research because I am interested in discovering what is unknown," Brian answered.

The professor replied, "Yes, I know. But I have seen too many bright kids not like research. Not that there is anything wrong about that. I suggest you do work in a lab during your senior year and see if you like it. From my own experience, I can tell you tha nothing is more exciting than running a successful experiment in the laboratory. However, as you will see for yourself, the majority of experiments do not work as expected, so you have to repeat them over and over again. For a doctor it is not like that, as you can imagine. You will need to diagnose a patient the first time you see them. Therefore, it is important that you shadow a great doctor as they see patients, make diagnoses, and prescribe medications. So you must prepare, take the MCAT and apply to schools. All major research-intensive universities offer a MD-Ph.D. degree. That would be your ticket to greatness."

Brian enjoyed shadowing doctors and working in the micro-biology lab of Dr. Janet Finston, studying antibiotic resistance. He scored high enough to be enrolled in a prestigious medical school in New York for a joint MD-Ph.D. program. He thought music could be his secondary passion. However, instead of going to classes, he again started wandering over to the New York School of Music to play the piano. He failed his first anatomy class. With a call from his mother and reminder of his brother's condition, he started paying more attention to his studies and less to music. Finally, after 10 years, he earned both a medical degree and a Ph.D. in the biochemical basis of protein synthesis. He now felt that he had developed as a physician-scientist.

Realizing that a full time profession as a physician may not be enough to make a huge impact to cure genetic diseases like the one that affected his brother, Brian started visiting various top laboratories to do his postdoctoral work, honing the skills of genetic manipulations that he thought would be necessary for him to perform independent research on the genetic basis of human diseases. He was looking for something that would grab his interest. He found it in the lab of Dr. Marc Levin, a renowned yeast geneticist at a state university in upstate New York. Dr. Levin was performing gene manipulation in yeast. He was doing various genetic tricks such as recombination (joining various genes), gene replacement and gene mutation studies to understand how genes function or how biology works.

Now, Brian had found out what he was looking for, and that he could be a creative scientist. He did not see much difference between playing music and in conducting scientific experiments. Both seemed somewhat repetitive and tedious, but at the end one gets great satisfaction from successful results. Research is not that different from sitting in front of a piano to master a piece by practicing over and over. To be a master of anything, you have to do it over and over. His undergraduate professor was right.

He finished his postgraduate studies with Dr. Levin and started his own laboratory in a reputed university in Maryland, to continue what he had learned in yeast and apply it to mice, a species genetically closer to humans. He had a perfect combination, a medical degree and experience in genetic research.

Brian also always wanted to collaborate with scientists working in a new exciting field that would make an impact on society. He was always looking to get deeper into this field. In school, he learned about gene therapy which he realized had huge promise to cure genetic diseases. However, for quite some time, this field had its ups and downs. Scientists would see phenomenal results of gene therapy with animals, particularly in mice, but the treatments failed miserably when applied to humans. Two major obstacles were found. The first one was that the introduced genes would function properly for a short period of time before being inactivated by the host cell machinery. The other was that the cellular defense, the immune system, would reject the new gene outright, thinking it was a foreign invader.

However, there were successes. A rare immune disease called severe combined immunodeficiency (SCID) was partially cured in a 4 year old girl. As the name suggests, because of a highly impaired immune system, SCID patients are extremely vulnerable to various kinds of infectious agents. SCID is actually a group of several different diseases due to defects in one of several genes. However, in all cases, one type of blood cells, T cells (white blood cells that identify and attack foreign invaders), become non-functional and thus, another type of cells called B cells (white blood cells that produce antibodies against infection) cannot produce the antibody against that foreign invader. In this case, the affected girl had a dysfunctional enzyme called adenosine deaminase (ADA) in her blood cells. Her blood cells were isolated and a healthy version of the ADA gene was inserted into a retrovirus (a virus with a RNA genome) to put it into the white blood cells. The modified cells were infused back into her body. Four years later, almost half of the girl's white blood cells carried the healthy gene.

Then, there was the incidence of hemophilia, a blood clotting disorder caused by a damaged gene, clotting factor IX. Scientists on both the East and West coasts cured hemophilia in dogs by introducing a healthy factor IX gene using adeno-associated virus, a new virus vehicle.

Unfortunately, there was a huge setback following a human trial to cure a rare metabolic disease. An 18 year-old patient died as result of gene therapy in a Philadelphia clinic.

To start on his own, Brian wrote a grant application at a governmental agency to develop a mouse model to study a method to cure ALD in the US, the disease his brother had. Brian was granted a multi-year study grant. He learned that a group had already developed a mouse model for ALD. So, he consulted with the authors about his plan to develop a similar model in a different type of mouse with new procedures. After long and labor-intensive studies, he finally succeeded in developing a new mouse model. He confirmed the earlier observation that X-linked ALD mice exhibited similar biochemical defects to those found in human X-linked ALD and published the results. The results provided an experimental system for testing therapeutic intervention.

Brian's group then tried to see if gene therapy could succeed in his mouse model. He knew a group in France who pioneered a new lentivirus technology and they applied it on an ALD mouse model. Before this technology, the treatment for ALD relied on bone marrow transplantation, an approach limited by the scarcity of donors and the risk of serious complications. It was known that lentiviruses, such as HIV-1, have a deceiving power to slip into human genes without activating the immune system. Brian's group thought that they could utilize lentivirus to introduce a healthy ABCD1 gene into mice. Now, they needed to find a place to put the gene. It was also known that bone marrow stem cells are capable of generating various cells-types of the body. They realized that if they extracted bone marrow stem cells (also known as hematopoietic stem cells, or HSC, cells) from a mouse, modified them, and then re-introduced the HSC cells into the same mouse, then there should not be a problem of rejection or other complications the earlier investigators encountered.

For this procedure, the HSC cells from a mouse were harvested and the healthy ABCD1 gene was inserted into the cells using a modified lentivirus. Finally, the modified HSC cells with the healthy gene were transplanted back into the mouse. They hoped that some of these cells would find their way to the mouse brain where they could express the healthy ABCD1 gene and cure the ALD. The result was encouraging, as the treatment resulted in almost a quarter of brain microglial cells producing the ALD protein 12 months after transplantation.

The next logical step was a study in real patients. The French group had already initiated a gene therapy trial in two ALD patients. They removed the HSCs from the patients, genetically corrected them outside the body (*ex vivo*) in a test tube with a lentiviral vector encoding a healthy ABCD1 gene, and then reintroduced the cells into the patients in a process very similar to what they had done in mice. They detected ALD protein in many different cells types after several months of follow up. Up to 14% of blood cells called granulocytes, monocytes, T and B lymphocytes made the ALD protein, leading to progressive cerebral recovery in two patients. These results strongly suggested that the modified HSCs were successfully transduced in the patients. Thus, lentiviral-mediated gene therapy of hematopoietic stem cells succeeded in providing clinical benefits in ALD.

With that knowledge, Brian wanted to initiate ALD therapy in the US. He was devastated when he realized that his brother Roger could not participate in this new gene therapy or the earlier hematopoietic cell transplantation studies because of excessive progression of his disease. Brian was depressed, but not discouraged, and proceeded with studies in younger patients. He recruited two young ALD patients. However, he wanted to try a new trick. In this case he wanted to introduce the healthy ABCD1 gene in a very specific place in the chromosome. One of the problems with lentiviral procedure is that lentivirus has a tendency to home in random places, which could have disastrous consequences if it disrupted an essential gene. To avoid this problem, Brian wanted to have the gene be introduced at a predetermined place. He experimented this new procedure in mice and it worked extremely well. This procedure utilized a designer enzyme, called a Zn-finger nuclease which would cut the DNA at a specific location, and determine where the ABCD1 gene gets inserted.

So, he tried this procedure in his human trial. Brian removed HSC cells from the patients and inserted the healthy ABCD1 gene with extra DNA sequences on either side, matching the sequences in the chromosome where he wanted the ABCD1 to insert. He then treated the HSC with the modified ABCD1 gene and the designer Zn-finger nuclease. The modified HSCs with healthy gene were isolated, grown and reintroduced into the patients. After several months of follow up, he observed definite progress of the patients similar to the observations of the French group. He published his results in a highly reputed journal. With this significant advance, Dr. Brian Warten became internationally known as one of the best human gene therapists in the US.

A letter from Germany brought sad news to Frederick's home. His father had passed away from colon cancer. As a physician-researcher, Brian was very upset, especially because he had visited his grandparents last year without much indication of the seriousness of the disease and the fact that he couldn't do anything. He was aware from his training in medical school that certain cancers can run in the family. However, most colorectal cancers occur in people without a family history of such cancer. Still, as many as 1 in 5 people who develop colorectal cancer may have other family members who have been affected by this disease. So, without taking any chances he asked his parents, who were in their early sixties, to have colonoscopies done, as this is one of the best ways to catch the disease before it's too late. Good news: no polyps. All clear for now.

With two different types of diseases in his family, and realizing enormous promise, Brian became focused entirely on gene therapy. He received a call from one of his colleague friends, Dr. Sidney Bock from Houston, who was also a great gene therapist and a pioneer developing gene therapy to cure blindness.

"Hello, this is Brian Warten."

"Oh, hello Brian, this is Dr. Bock. How are you?" came the response from the other end. Dr. Bock just finished his trial on blindness. It was on a rare inherited disease called Leber's Congenital Amaurosis (LCA), in which a missing protein, due to mutation of any of 13 known genes in human mitochondrial DNA, damages cells in the retina, the light sensing film that lines the back of the eye, and causes blindness.

"How did the trial go?" Brian asked.

"Oh, I just couldn't believe it, the poor girl could walk through an obstacle course without any help," came the excited response from Dr. Bock.

"Oh, this is outstanding news," said Brian. "What's your next plan?"

"We are going to start a larger trial with some younger and older patients," answered Dr. Bock.

"So, tell me, what vector did you use for this trial? I knew you are a fan of AAV (adeno associated vector) or lentivirus vector," said Brian.

"For this trial, we used AAV. It worked great," answered Dr. Bock. "Oh, can you please hold Brian, don't hang up. I need to take this call," said Dr. Bock. Dr. Bock returned after two minutes.

"Guess what Brian? Can you guess who just called me?" asked Dr. Bock.

"Oh, I wish I was a mind reader, Dr. Bock. Your wife, the high command?" answered Brian.

"No. The mother of the patient who had the successful trial called me to say that her daughter is now complaining that the sun is too much for her to go out. She is overjoyed that this is the first time she complained about light," says Dr. Bock.

"That's pretty awesome, Dr. Bock. 12 years in the dark, what can you say? Congratulations on your successful therapy," Brian replied.

"Thanks. So, what's next for you, Brian?" asked Dr. Bock.

"I will let you know in the next few weeks. Something is cooking, but not enough information yet," says Brian. "Fair enough, Brian. Okay, please keep me in the loop if you like. Thank you, Brian. Okay I need to call the mother back and find out some more information," says Dr. Bock.

"Of course, I will. Okay, talk to you later. Bye." Brian hung up the phone.

Brian couldn't be more excited to join the field with full steam. Back in his mind was the death of his grandfather who had succumbed to colon cancer. He wanted to do something to cure cancer. Could gene therapy be the answer? Cancer is generally a disease of old age. However, younger people are also affected. In addition, cancer is a complex disease. There are more than 200 types of cancer. So, there may not be a single treatment for all cancer. Most cancer therapies that are successful are chemotherapy and activation T-cell mediated immune therapy. Both have their own drawbacks, side effects in particular. So, Brian was looking for something special, something simple that could be used as a universal treatment for cancer. It's a tall order, but why not think big?

While reading an article in a scientific journal, he noticed an announcement that there will be a conference in Washington D.C. next month on the patenting of genes. He was busy with so many things that he totally forgot that the conference committee had invited him to deliver a keynote speech on his ALD gene therapy work. So, before forgetting again, he replied by e-mail to accept the invitation.

From the agenda of the meeting, he found an interesting study from Chicago to be presented in the conference. He knew from one of his friends that the investigator, Dr. Sam Roy, had found a bacterial protein that preferentially invades cancer cells and kills them by multiple mechanisms. However, he did not know if that protein has been tried in patients. He hoped to meet the investigator and learn as much as possible about the protein. This could be the protein he was looking for a long time and could be tested in gene therapy trials for cancer.

The conference was well attended and there were many notable scientists who spoke. Brian's talk was very well received. Brian was also present at Sam Roy's talk and decided to have a chat with him afterwards. Following his talk Brian wanted to introduce himself to Sam Roy. However, he was so busy talking with the journalists that he didn't get a chance to talk to him. So, he came back to the hotel and left a note with the hotel receptionist and asked her to hand it over to Dr. Roy who was also staying in the same hotel. Sam had already emailed Brian a week ago and made an appointment.

A few hours later, Sam Roy called Brian Warten and invited him to dinner. At the dinner, Brian Warten and Sam Roy met for the first time. Sam wanted to tell him about the grand challenge he had received from Bluefrostland and wondered if Brian would be interested in a collaborative project with Sam.

seventeen
The grand challenge from the Bluefrostland Ministry of Health

As Sam's animal experiments started to show highly promising results on the anticancer and cancer preventive activity in mice, and he received lot of publicity, he began to get a large number of inquiries from many people, particularly patients and relatives of patients with advanced cancer, on the possibility of trying rhabdosin in them.

One of the most striking letters was from the health secretary of the Ministry of Health of Bluefrostland, an island located northeast of Greenland and Iceland. Nobody knew much about this island other than some brave souls who liked cold weather, skiing and ice fishing, and sun worship during the summer when the sun was always up. Then a few years ago, a spectacular thing happened; the island found vast amount of oil under its sea bed which attracted attention of the major oil producing companies from Europe and the US.

The economy of the Bluefrostland hit its zenith, attracting not only oil and gas workers, but also many real estate developers seeking to build a mirror image of some of the equally prosperous Middle Eastern countries. The tiny government of Bluefrostland was initially overwhelmed but soon realized the possibilities of attracting well-educated and business-oriented immigrants. They decided to use their oil wealth to build an island country with the finest tourist villages on the coast with the best medical care to attract people from all over the world. They wanted to be the next big destination for scientists, engineers, doctors and thought-leaders from all over the world. They decided scientific facts would drive the national priorities, not religion, not popular opinions, and of course not politics. They wanted to be the most progressive country in the world and cutting edge science pushing the boundary of knowledge, would lead the way.

One of their strategies was to introduce the concept of grand challenges in medical and health care whereby Bluefrostland invited premier technologists, medical researchers, physicians and marine biologists to explore the country's unique Arctic marine flora and fauna to develop new kinds of medicines, drugs and medical care. Aware of Sam's highly publicized development of anticancer and cancer preventive bacterial protein drugs, the Bluefrostland Ministry of Health extended an invitation to Sam to set up his laboratory for new kinds of anticancer drug development from marine bacteria unique to the Arctic climate. The island nation would provide not only the finest laboratory facility that money could buy, but also the facilities for clinical trials in well-equipped hospitals. The offer emphasized that although the island would own the intellectual property generated through such activities, they would reward the researchers and clinicians with significant royalty payments. Sam was amazed. The letter couldn't have come at a better time. He talked to many of his colleagues, who were all excited to be a part of such effort and spending summers (and dark winter months) in an island that could be heavenly. He also had to let Erin know about the opportunity. He was worried about how this was going to affect their lives.

After an exchange of letters to define all the facilities, particularly the clinical facility for trials with cancer patients who could move from Europe or the US to get treated with the new drugs in

Bluefrostland, Sam signed a five year multi-million dollar contract. He was confident that from the coastal marine microorganisms of Bluefrostland he would be able to find a few unique bacterial proteins to use as anticancer and cancer preventive drugs. He spoke to some of his potential patients, who were meant to be treated with rhabdosin once he obtained regulatory approval, to move to Bluefrostland, where he would treat them free and with substantial compensation with any new drug developed in Bluefrostland, and with full freedom to accept or deny treatment anytime they choose to. He also decided to talk to Brian during the upcoming meeting in Washington D.C., for his willingness to conduct the gene therapy trials in Bluefrostland.

eighteen

Brian and Sam meet: The meeting of minds

It's very nice meeting you, Dr. Warten. I have known about your work, but have not met you in person before," said Dr. Sam Roy.

"Thank you, Dr. Roy. Please call me Brian. And if you don't mind, I will call you Sam. Yes, I knew something about your work before, and now, I know a lot more after your seminar," replied Brian.

"No problem, Brian. You can call me Sam," replied Sam.

Before Brian could say anything, Sam congratulated Brian, "Your ALD gene therapy work was just outstanding. How are the patients doing right now?"

"One of the patients is doing very well. The disease progression has completely stopped. The other patient is doing great as well as it slowed down the disease progression considerably, but we may need to follow up in next 6 months or so to find out if the progression stopped completely like the other patient," Brian replied.

"That sounds great, Brian. I am here with a proposal for you," said Sam.

"Proposal? What is it?" Brian was curious but could not guess what was in Sam's mind.

"You know our results with rhabdosin, the protein made by the bacteria *Rhabdosis pulmoneria*. As an infectious disease specialist, I have been working on this bacterium for many years. First we showed that it was effective in shrinking tumors in mice. It can even prevent cancer in lab experiments in mice. But I know that one company has tested a peptide called p28 in 15 patients with cancer, and found that it has no toxicity. Meanwhile, I had permission from our institutional research board to try to get approval from the FDA. I have 5 patients with breast or ovarian cancer who we plan on injecting with rhabdosin three times a week after we get FDA approval," said Sam.

"I see. So, I suppose you are trying to propose to try gene therapy on them, right?" Brian smiled.

"Yes, of course! But it's more than that. A few weeks back I got a letter from the ministry of health of an island way northeast of the Arctic Circle named Bluefrostland. They found a lot of oil under their seabed and became rich. They are, however, sparsely populated and want to attract talented immigrants with new visions. I am not claiming that I have that, but they sent me a letter offering all the facilities we might need to conduct both drug development and gene therapy trials. I am motivated by the once-in-a-life time opportunity to do what I am dreaming of doing. I want you to be a partner in this effort. Needless to say, you will be well compensated. My idea is to develop a new bacterial protein from the marine bacteria and put the bacterial gene encoding this protein with anticancer activity in cancer patients. Do you think it will work?"

"Putting a bacterial gene in the human genome is a new concept that most people will be anxious about. I know that our guts are full of bacteria and many of them are very useful for our health and well-being. There are extensive reports as to how important the microbiome is for how we digest food, and stay well, but a bacterial gene inside human cells? I don't know how that will go! But, your proposal is both intriguing and exciting. " Brian exclaimed.

Brian held his head in his hands, closed his eyes and thought for a moment before continuing with the conversation. "Gene therapy at present is very much an experimental therapy. Previous trials did not work well because of the premature nature of the scientific advancement at that time. It gave people a lot of hope, only to show that it was not ready for prime time. It was just not the right time. However, an early clinical trial looks promising. I just read a paper about the effects of introducing a gene known as factor IX in hemophilia patients. These patients do not have this clotting factor and therefore are in danger of bleeding to death. The paper reported that six such patients were injected via peripheral vein infusion with the factor IX gene carried by an adeno-associated virus. After the injections, some of the patients had improved bleeding times with very few side effects, and more patients are being recruited. Our experiments with ALD patients have also met with partial success. I think that, like the factor IX gene, anticancer protein can also be made from genes we will isolate in Bluefrostland. Of course, I have to do more experiments before we can begin. Since any such gene is absent in the human genome, we have to be cautious in looking for any undesirable side effects. We have to know if the protein will be soluble and modified inside the body to be active." As Brian was explaining, the waiter was waiting patiently for their orders. Sam ordered tandoori chicken for both of them. This hotel had an Indian restaurant and there were several Indian dishes on the menu. Brian had never had Indian food before, but he was willing to try. He felt more adventurous and became curious about the research and clinical trial possibilities in an Arctic island.

"So, what do you think? Will it be possible to do in a human? As I said before, I have a patient who is willing to try," Sam asked.

"Oh, what about the immune response? Being a bacterial protein, it might induce immune response. Or, what about antibodies against the protein?" Brian asked.

"I am glad you asked those questions. Usually these proteins are totally non-immunogenic. The structure of the protein resembles human antibodies. So, the body somehow does not see such proteins as foreign proteins. But we will not know about any immunogenicity or toxicity until we isolate such proteins in Blue-frostland and try them out."

"That's true. So, a protein like this could have a perfect application in gene therapy if there is no immune response and no toxicity. We can go first for somatic gene therapy similar to what we have done for ALD. What do you think?" asked Brian.

"Yes, that's great. But, have you ever thought about germ line therapy?" Sam asked.

"Of course, but we are not there yet. The technologies are there, but at this time it's too risky and expensive. Just in case you want to know how complicated the procedure is, here it is in short, although I may not remember fully all the steps. Germ line therapy is a multistep procedure and several things can go wrong. But, if we try, I would prefer using *in vitro* fertilization for this therapy. First, we would need to isolate embryonic cells at an undifferentiated stage. Then, we would need to grow embryonic stem cells in culture. More cells would be needed for gene targeting. We would then transfer the bacterial protein gene into embryonic cells. Since millions of cells must be transfected to obtain even one targeted recombinant, mass transfection techniques would have to be employed. As I would prefer to insert the gene in a specific place in the chromosome, the bacterial protein gene would be flanked by extra DNA sequences on either end. The extra sequences would target the insertion to the intended location. The cells will be transfected with Zn-finger nuclease (it could be TALEN as well) designed to cut a particular sequence at the point of insertion, along with the bacterial protein gene cassette, encouraging integration at the targeted location. The selection of cells which have stably taken up the bacterial gene would be done by a special procedure. I don't think we need to go into detail on this at this time. Once we have cells with the bacterial protein gene, unfertilized ova would need to be recovered from a super-ovulated woman. The nucleus from ova would need to be removed and replaced with a nucleus from our new cells with the bacterial protein. An embryo could grow, carrying the modified genome. Then it would be re-implanted into the woman. Now, the woman should be warned that (1) there could be spontaneous abortion or miscarriage; (2) only about 15% of *in vitro* fertilizations lead to successful pregnancy; (3) to increase the chance of success, multiple embryos would need to be re-implanted simultaneously, and (4) if successful, identical twins or triplets may result. Also, there might be legal and ethical issues that we are not in a position to handle now but for which we would have to consult with people and government in the island. I am very optimistic that the regulatory body of the island will allow us to do it because it is based on scientific facts. Can something go wrong? Of course. But we will take all precautions. It's like what NASA did with sending people in the orbit. Could they guarantee that it was completely safe? No, but what they had suggested that it was worth taking the risk as future benefits are countless. Did some of the missions fail? Of course, the Columbia shuttle burned as it entered the earth atmosphere. We learned from it, and made safer shuttles. I don't want to compare these two frontiers, but my point is that here is an opportunity for us as scientists to push this boundary, and this island provides us or at least promised to provide us with such an opportunity. If we can save lives for the future generations, this ethical issue hopefully will be a moot point. It will not only save lives, but it will also save the medical costs which are now going out of control," Brian explained.

"I am so glad that you see it that way, Brian. Someone has to start first. So, what are you thinking?" Sam asked.

"I think the best way would be to start with somatic gene therapy, which has a better chance of success. We will follow the patients for about 6 months. If we see progress we can start germ line therapy later," said Brian.

"Sounds like a good plan," said Sam.

"But, what about the patients? I will need to speak to the patients. Do you have anyone in mind, Sam?" asked Brian.

"Yes, I have a 28 year-old female patient who has mutations in the BRCA1 gene. She has ovarian cancer. Her mother had breast cancer. She volunteered to be the first subject of this gene therapy trial. She is afraid that because of her family history, she would pass the mutations to her children. She does not want to have her ovaries removed. She wants to have children. She is also afraid she will have a recurrence a few years later," said Sam.

"Okay, we can give it a try. But, as I said before, I would like to try somatic therapy first and watch progress before we even think about germ line therapy. Also, as said earlier, I prefer to use *in vitro* fertilization to start germ line therapy, if we ever choose to do it. Unless she gets married soon, in which case we can wait more than 6 months. I hope you will agree," Brian said confidently.

"Certainly, I will have to ask the patient about her plan. In any case, I would like to introduce her to you in the next couple of days. You can educate her on the pros and cons of the procedure and expected and unexpected outcome," replied Sam.

"Yes, no problem," said Brian. "The chicken tastes good. Sam, because of you, I am becoming addicted to Indian food. Although my wife says I am very rigid in my food habits. What about you, Sam? Are you married?" Brian asked.

"Not yet! There is a chance that I may have found a future wife. She will be jealous, though. She says I am married to my work," Sam answered. They had a great dinner. Brian and Sam walked up to the hotel conversing casually. They felt like old friends. Brian gave Sam his telephone number so that he and the patient could discuss the future gene therapy experimental trials.

nineteen

Preparatory discussions and plans for action

A phone rang in Brian's office. On the other end were Sam and Kathy. They wanted to discuss the plan of gene therapy. Brian was not in his office at that time. So, the phone was automatically directed to his secretary.

Brian's secretary picked up the phone, "Hello, this is Dr. Warten's Office. May I help you?"

Sam on the other end said, "Yes, may I speak to Dr. Brian Warten? I am Dr. Sam Roy from Chicago."

"Sorry, he is not in his office at this time. He is making his rounds and seeing his patients. But, I expect him to come back in about 10 to 15 minutes. Would you mind giving your contact number so that he can call you when he returns, and do you have any specific message?" asked the secretary.

"Oh, sure! Please tell him that Sam Roy and his patient called from Chicago. Dr. Warten is aware of the topic we will be discussing," Sam explained.

"Okay, thank you," she said and hung up the phone.

Brian came back shortly after the phone. He got the message from his secretary. Then he called Sam.

"Hello. This is Sam," Sam answered.

"Oh, hello Sam. It's Brian. Sorry, I missed your call."

"No problem, I called to introduce Kathy, the patient I mentioned to you in D.C. You mentioned that you want to talk to her before going any further. So, here she is. Do you want to talk to her now?" Sam asked.

"Oh, that's great. I could talk to her privately, but I think it would be great if you could also join in the discussion. I just wanted to explain to her about some of the procedures we will be doing and talk with her about the expected and unexpected outcomes. Your presence may give her more confidence," Brian responded.

Sam turned on the speaker phone and closed the door. "Brian, now you are on the speaker phone," Sam told Brian.

"Thanks, great," Brian responded.

"Hello, Dr. Warten, this is Kathy. How are you?" Kathy said.

"Hello Kathy. I am fine, thank you. Kathy, I am sure Sam has already informed you about the plan of having somatic gene therapy, right?" Brian answered.

"Yes, he did," Kathy responded. Kathy was very nervous and was not sure what he would say.

"Okay, good. I will just go over with you the procedure and will try to educate you so that you will feel more comfortable with the entire process. How does it sound?" Brian asked Kathy.

"Okay, it sounds good. Thank you." She was still very nervous.

"The procedure we will do is called 'somatic cell gene therapy.' It means that in this treatment only you will be cured if everything goes well as planned. It will not pass to the next generation. As, Sam may have already told you, we will modify your blood cells to introduce the gene for a bacterial protein into your chromosomes. Right now, your are being treated with conventional drugs. However, we are planning to go to an Arctic island, called Bluefrostland. We plan to do more research and may have bacterial proteins that are more suitable than the conventional drugs. Now, here is the thing. We cannot use existing blood cells; we will have to collect some of your bone marrow, which contain blood stem cells called hematopoietic stem cells. We call them HSCs. These cells are responsible for making all different types of blood cells. If we introduce a bacterial protein gene in HSCs, all blood cells will produce this protein and will hopefully protect you for life. Do you follow, Kathy?"

She answered, "Yes, I follow."

Brian continued, "However, bone marrow contains other types of cells as well. Once we take your bone marrow, we need to separate true HSCs from other types of cells. You don't need to worry about it because we are the experts in separating the HSCs from other cells. HSCs are not present in huge amounts. So, the next step will be to grow them in the laboratory so that we will have enough HSCs to work with. Now the most important step—introducing the anticancer bacterial protein gene into the HSC chromosomes."

"I am getting a bit scared, Dr. Warten."

"Don't worry, Kathy. You are in good hands. We have performed a similar procedure with other patients. Although they are not cancer patients, they are doing great. Believe me, the procedure works. Okay, where was I? Now, how do we introduce the anticancer protein gene into HSC chromosome? You know, we took advantage of the technique some type of viruses use. We have a vehicle which comes from a lentivirus, like HIV-1, and they can slip their genes into human chromosome," Brian continued.

"Please wait Dr. Warten. I am not going to go through this. To cure one disease, I don't want to get another disease or, may be both, if your treatment for cancer does not work," Kathy told the doctor.

Sam tried to calm her down. "Don't worry, Kathy. That's not the full story. Please let him finish; and then you see if it is bad, Okay?"

"Thank you, Sam. Kathy, please don't worry. We will not do anything unless we are sure it's going to work without doing any harm. We are actually using a completely disabled virus, so don't worry. We took out the harmful genes of the virus and put only the gene we want to introduce and in your case it will be the gene for a protein drug. So, we will treat the HSCs with this modified virus vehicle in a special way. Once the modified virus infects HSCs, some of the cells will take the DNA and insert it into the chromosome. And, since the virus is disabled, it cannot make active virus. Are you with me, Kathy?"

"Yes, so far. But, where does it go in the chromosome, Dr. Warten? Do they go in a specific place?" Kathy asked.

Brian responded, "Oh my God, you are extremely intelligent. Yes, although it has some hot spots, the site can be random. So, there may be some potential problems as it can inactivate some good genes. But, we have a strategy to solve the problem. We have devised a strategy. It is called Zn-finger technology, so that we know exactly where we want the gene to go. Are you feeling better, Kathy?"

"Yes, much better," Kathy answered.

"See, Kathy, I told you before, once you hear the complete story you will not have to worry about it," Sam assured her. I have just one more point to add. These new technologies are experimental, extremely costly and must be done in a country that is rich enough to provide all the facilities. Unfortunately, this is not in the US, but in an island north of the Arctic circle. It's beautiful though, particularly in the summer. Its name is Bluefrostland and both Sam and I will spend a few years trying out our new anticancer and cancer preventive drugs. Will you come?

"Sure, but how much will it cost?"

Sam followed "Kathy, no cost at all. This island is wealthy and they will cover all your costs and compensate you financially. So please don't worry about it."

"Now let's go to the final steps. We need to do some tests to make sure that the gene for the protein drug is integrated in the HSC chromosome. Once that is done, we will separate them from the unmodified HSCs. Again, the modified HSCs will be expanded as we did earlier. Then these modified HSCs will be reintroduced into your body. Then for several months we will follow your blood for the presence of the protein drug. There is a greater than 90% chance that it will be present. As you can imagine, blood goes almost everywhere. Cancer cells make more blood vessels because they need more nutrients to survive and grow rapidly. So, wherever blood goes, the drug will go with it. When blood goes to cancer cells, the drug will invade and kill them. That will be a huge success story, won't it?" Brian explained.

"Of course, Dr. Warten this is wonderful. I can't wait any more. I hope my initial treatment is in the summer months so that I can take an occasional walk along the beach" Kathy replied.

"We will have to do many preliminary experiments after we arrive in the island. We will have to do toxicity studies with volunteers. It may be a few more months. Much depends upon what we find once we are there. Is it okay Sam? Or, do you want to discuss further?" asked Brian.

"No, that sounds perfect. I am glad that you took time to discuss it in detail with Kathy. She is now very confident with the procedure. We will do it," answered Sam.

Before Sam decided to accept the grand challenge offer, he took a one-week trip to Bluefrostland to finalize his negotiations. What he saw was impressive. The coast in the summer was beautiful with occasional small icebergs floating by. Seals, dolphins and many other aquatic animals, but not any sharks, were plentiful and playful, but the most spectacular thing was the new medical center campus where his unit would be located.

It was packed with many pieces of fancy equipment, and his first year budget was 30 million dollars, to allow him to buy other pieces of equipment as he needed them, and to hire about 50 professionals, medical students, interns and postdoctoral fellows.

There was a separate hospital where his partner Brian would be located, with safety cabinets under positive and negative pressures to allow him to conduct basic research on gene therapy. Brian would have a separate budget of $25 million per year to conduct clinical trials.

Sam could not have asked for more. He signed the contract for himself and brought a copy for Brian to take a look at his own budget and laboratory/clinical facilities. When he returned to the US, he was jet-lagged but happy as a clam. He decided to return to Bluefrostland in six months after obtaining a sabbatical leave from his university.

After they decided to collaborate, Brian and Sam met again to discuss their experiments.

Brian started the conversation. "Sam, before I approach the institutional research board and the Bluefrostland government regulatory agency for approval of the testing of our new drug in Kathy, I have to supply them with data from detailed studies on mice or other animals so that they are convinced that the drug has no toxicity and good efficacy for cancer therapy. So please tell me about your proposed initial experiments with mice or how you want to go about it."

Sam replied, "Brian, I have no idea what kind of candidate drug we may come up with in Bluefrostland, but I am very hopeful that the marine bacteria, our beloved bugs, from the coastal areas of Bluefrostland, will produce smart protein weapons, similar to rhabdosin, targeted towards cancer. So my approach will be to isolate marine bacteria from the large aquatic animals, dolphins, seals, even sharks, where such bacteria colonize these animals for a symbiotic relationship and consider them as their habitat. So it is natural that they will produce protein weapons to defend their hosts. I will thus do the same experiments with these protein weapons as I did with rhabdosin.

I have done two kinds of experiments with a large number of mice to get good statistical data with rhabdosin. To find out if our new drug will show toxicity and/or cancer regression, I will take 20 immuno-deficient mice and give them all subcutaneous injections of two human cancer cell lines: melanoma (skin cancer) and breast cancer. Because these mice, called nude mice, have no immunity, the human cancer cells will start to grow to produce tumors. After a few days, when all the mice will have tiny tumors visible under the skin, I will divide them in two groups of 10 mice each. To one group, I will give injections of sterile saline three times a week. This will be the control group. To the other group of 10 mice, called the test group, I will give injections of 1.0 mg of the candidate drug 3 times a week in the belly, away from the tumors so that the drug has to reach the tumors through the circulated blood. Then, twice every week, I will measure the size of the tumors to see if the drug-treated mice have smaller tumors than the non-treated mice. As you can imagine, I expect to see after 4 or 5 weeks that the drug-treated test mice will have tumors 50 to 80% smaller than the tumors of the control group. I also will sacrifice some of these control and test mice, isolate the tumors and study them histologically. The studies will likely show that the drug-treated tumors will be very soft as opposed to the solid tumors in the control mice, suggesting that the drug kills the tumor cells by allowing them to soften and burst. In addition, it is likely that the drug-treated mice will have no visible signs of toxicity, good mobility and will gain weight, as we have shown for rhabdosin. These will be our initial approach. We will also look for antibody formation against the protein drug in mice to ensure that our drug is non-immunogenic and does not elicit an immune response in these mice."

"This is very interesting." Brian was absorbing the information. "What will be the other experiment as you have for rhabdosin?"

Sam replied "I am also curious to see if our new protein drug will be able to prevent cancer rather than just kill pre-formed tumors. For that, I will take 20 mice. All will be fed the carcinogen 4-nitroquinoline-1-oxide (NQO) with their drinking water. NQO is known to induce oral and esophageal cancer in mice when given orally for several weeks. In our control group of NQO-fed mice, I will not give injections of our drug, while in the test group of 10 mice, I will give injections of 1.0 mg drug every alternate day for several weeks and will then count the number of lesions (pre-cancerous) and tiny tumors. I hope there will be very few lesions and essentially no tiny tumors in our drug-treated NQO-fed mice. Such results will show if our drug may prevent the onset of cancer in carcinogen-fed mice. I have conducted similar experiments with rhabdosin using a different carcinogen dimethyl-benz-anthracene and I know how to test the cancer preventive activity of any new drug we find in Bluefrostland."

"This is great!" said Brian. "Now, Sam, I know that you have progressed far beyond those first experiments with rhabdosin and I am as enthusiastic as you are about the possibilities with our new candidate drugs in the island. For our gene therapy trials to proceed, you have to do studies with mice, to show that the protein drug gene can be introduced into the stem cells of mice."

"Yes," said Sam. "I will call a lab meeting and get started on it. Please send me all the relevant references. I do not know how many of my present lab members might be willing to accompany me to Bluefrostland, but recruiting competent researchers in the island should not be a problem."

twenty
Clinical trials in Bluefrostland

S am arrived in Bluefrostland on a late spring evening, and there was sunlight everywhere. His first few days were hectic, but his mind was occupied with a frustrating sense of missing Erin's company. He hadn't seen her since the Hawaii vacation and he was desperate to see and hug her. For her part, Erin was somewhat ambivalent about missing Sam. As Erin's vacation had ended on a very disturbing note, she wanted badly to sort things out, but just didn't know how. She devoted herself wholeheartedly to her work and fund raising projects but try as she might couldn't break up completely with Sam. She kept her correspondence with Sam limited to email exchanges. She vowed to maintain her distance so that she could get a better grasp on her emotions and needs and also to give Sam space to realize what mattered to him.

Their Hawaii vacation, flawed as it had been with Sam's thoughts always going back to his endless scientific pursuits, was the nearest thing to perfection that she had ever known. Sam's care and deep understanding of her innermost desires and her health issues always took her by surprise, specially because they had grown up on practically opposite sides of the globe. To her that was another reason why their love was so special so rare and hence invaluable. She was positive Sam felt the same way. His beautiful poem that he had read out so emotionally in Hawaii echoed in her mind. It brought tears to her eyes to think how someone could make her feel so perfect when in truth she was flawed with some bad mutations. If his never-ending quest for knowledge always managed to make itself known at the most awkward of times, his honest acceptance of what was important to him just endeared him to Erin all the more. She continued writing on the poetry blog where her inner dilemmas and turmoil found a creative outlet. She posted this poem and hoped against hope that Sam would read it. It was called Moments.

I shall keep them
the moments—pristine and intact,
their language as rich with nuances and accents as they had been
when you spoke,
as alive with metaphors and metaphysics as they were
when you smiled,
like jewel fireflies in the tropical sunset,
fireflies you and I had seen and touched
with our wondering eyes and fingers,
spellbound together at the mystery of our natures;
you with your discoveries
I with mine,
sharing that perfect pitch and spark of give and take
only an instant maybe,
but a forever treasure,
before night came …

Beyond those moments
you are free to have your world,
a world I could surely steal had I wished so;
but
because of you, I know
some day, many shall learn to live anew, clean and healthy!
So if you crave tireless
to refresh and empower such a gasping and aching world,
how can I not wish you, my friend and path-finder, joy of the find?

in whatever way you may choose, where ever you may choose to go,
unconfused by my maps of predictable destinations,
unpolluted by the crosswinds of earthbound expectations?

The moments intact,
I cannot return or set free
even for you
they are mine and mine alone,
to live off, live for or live by
as I and I alone
shall desire.

She signed the poem 'Bonita,' and sure enough within a couple of days she saw Sam's reply in a poem called The Simple Truth.

The laughter, the tears, the moments, the years
are neither yours, nor mine,
but ours, golden girl,
the passion, the quest, the stumbles , the fears
will make sense
read the language of the sky and sea in your eyes
and bring the promise of healthy tomorrows in your veins
if I can feel your breath on my face when I gather my laurels …
We are in this together, golden girl
for just as you are bound by moments so am I
to transform the moments to forever …

It was signed 'Sam.'

Neither Erin nor Sam talked about their poems directly, but Sam kept asking her if she had made plans about visiting him. She debated if she should listen to his suggestion to visit him in Bluefrostland next spring. She dithered, toying with the idea, for she knew if she did indeed visit him, that would for sure be another step towards her commitment to him.

When Sam had left for Bluefrostland within a month of their return from Hawaii, on his insistence to get together before he left, she had just made a perfunctory visit to his lab but hadn't seen him off at the airport. Since then she received regular updates from him on how much freedom he had at work and how much he was enjoying the Arctic isle. But they did nothing to assuage her inner anguish. A war was still raging inside her. On the one hand, she couldn't get Sam out of her mind, on the other, her own life as she knew it meant much more to her than her feelings for him. At times she felt selfish when she chose the former, at times depressed when she opted for the latter. Finally she gave up exhausted and decided only time would tell. She did read up all the information she could find about Bluefrostland though, that it was made up of three separate islands of different sizes. The largest one was where there was the Central Mainland University hospital and research center, where Sam had headed as lead scientist.

And then, in early fall, Erin received an invitation from the Central Bluefrostland Library to give a talk at a conference on cultural integration to facilitate multi-talent inflow to the various science and technology fields. The conference was being hosted by the Central Bluefrostland Library near the Central Mainland University. She noted with pleasure that Dr. Sam Roy's name was also there as the main speaker in one of the sessions. The conference would be in early spring the following year just when Sam had asked her to visit him.

This was too much of a coincidence, and Erin couldn't help reading more into this turn of events. Did he have any role to play in her invitation? A year or so ago, she would have been enraged at this thought, or the slightest hint that someone had been pulling strings on her behalf. But she was guiltily surprised to find that she was rather relieved now that her decision had in a way been given an official tone. Whatever it was, she insisted to herself, this was just how things were playing out and she herself had nothing to do but go with the flow.

Erin remembered Sam's last mail that her visit to the isle was really necessary, for it would be very crucial in deciding the future turn of events in their lives. But his descriptions of the long winter months of cold and dark had depressed her beyond words. He was sweet enough to say that thoughts of Erin were like the Northern Lights or the Aurora Borealis that kept him going through all the trying times in a new land as he was trying to make a home where he had none. But Erin couldn't be sure how much credit she could take for it, although the very fact that he thought of her with passion and beauty couldn't help bringing a smile to her lips. And so it happened that more than her eagerness to attend the conference, Sam's boundless optimism in his newfound land, faith in his work and sheer positive energy finally convinced Erin to accept the invitation. She was also curious to know why Bluefrostland had attracted him so much. Above all she decided she wanted to give their relationship another chance.

When the plane touched down on Bluefrostland one late spring afternoon, the only thing different Erin noticed was that the airport was smaller and that there were fewer people around. They seemed unmistakably friendly and helpful. Not that she needed to ask much, for the airport was much more organized than she thought it would be, and there were signs in various languages with directions and information at every turn. Monorails ran right to the building and she had no hassle at all in finding the one to hotel. She had specifically asked Sam not to meet her at the airport in spite of his repeated insistence. She wanted to get a feel of the place without a biased companion by her side. For she knew that by meeting Sam after all this while she would be instantly carried away by his warmth and passion. Erin wanted to minimize the chances of being blown away by his optimism without getting the time to take in things in her very own way. After the three day conference she had taken another day off, for Sam had insisted he would take her in his new car to see the glaciers.

The hotel seemed a slightly modern take on the alpine log cottages she had visited on one of her European tours. Her room was on the third floor and one corner sloped down like a quaint cottage. She showered quickly and went to the lobby. The fireplaces were all lit, although it was spring. She ordered a coffee and was pleasantly surprised to find that her brew, though different from what she was used to, was quite strong and pleasantly fragrant. She found out that the beach was close by, as this was an island, and decided to take a walk right away in spite of her jet lag. There was a decided nip in the air outside, but it was different than anything she had known before. It was fresh and clean and had a faint smell of spruce. The lightness made her head tingle so much so that Erin thought of the word intoxicating. No wonder Sam loves it here, she made a mental note.

The main streets were wider than the side streets and they were all paved and pretty smooth. The cars were small and colorful and not too many. There were many bikers, but mostly pedestrians. The stores sold woolen garments, fishing, kayaking and skiing gear, but surprisingly no fur, and Erin saw signs forbidding hunting. She had texted Sam and said that she would meet him at his faculty office. As there were still a few hours for that, she walked on the beach enjoying the view. The waters were a true blue and the sands a pearly white, from which the island surely got its name. Otters and dolphins bobbed their heads here and there. There were snow covered mountains in the horizon and when Erin looked at the map she saw that the Silver Wonder, the highest peak and the surrounding evergreen forests were a popular camping and hiking spot. They were the focal point of the glaciers where Sam had promised to take her later.

Beautiful though Bluefrostland's online pictures were, Erin had come here with a lot of misgivings thinking she would see some kind of prehistoric land hostile to normal human habitation, specially to a city-bred girl like her. But she was rather taken aback at how familiar everything looked and felt. In fact she already liked the user friendly public monorail transport that whizzed her from the beach to Sam's research site and faculty building in barely fifteen minutes.

Sam was waiting outside the office building for her. There were some people going about their business, but it was easier in this Nordic land to spot him from far because of his dark hair and shorter stature. Sam must have spotted her too, for he turned from whoever he had been talking with and took long strides towards her. His lively hair swayed in the wind catching glints of sun and the closer he came Erin realized his hair had grown lighter, or did she see a few streaks of grey? She was waiting to see how her heart would react on seeing him again after all these months and as she was busy trying to analyze her thoughts, Sam came running up to her and she didn't have a chance to stop him from gathering her up in a warm, enveloping embrace.

For an instant she rested her head on his cozy shoulder, but as she quickly turned her face away, she heard Sam's easy laugh as he said teasingly, "Wow! So you really missed me that much! And I thought you had gotten me out of your mind and were probably planning to hook up elsewhere ..." he didn't get a chance to complete his sentence though for Erin pushed him away and looked at him with such outrage and disbelief that he embraced her again and whispered in her ear, "Erin I'm so happy you decided to come."

Sam had planned to take the rest of the afternoon off, and since the conference started pretty late the next day, Sam took Erin to his university housing. Erin found the two-story building like a modest size condo with large rooms and a very green backyard with tall spruce trees. The peace and majesty of the surroundings had a strange effect on her. She started to breathe more deeply and the ever observant Sam remarked, "Yes Erin, more than my work and what I want to achieve here it is this place itself that makes me more intense, inward, alive, if I may say so. It is so unpolluted that I can simply think better."

They had some grilled fish and potatoes that Sam had bought from the local fishery and farmer's market. They sat out on the glass windowed porch with their drinks and talked for a very long time. Suddenly Erin realized that it was almost 10 pm and the sun was not setting, but hung on the edge of horizon in a blaze of light like a still shot of a luminous sunset. Noticing her surprise Sam laughed and said "Remember how far up north we are? That's why I had wanted you to visit in spring so you could see that the sun never sets just like our …" he didn't finish but took Erin's hand and held it in his. Erin did not stop him and there they sat wrapped in a togetherness and deep need for each other, seeking the forever sun in each other's eyes.

The conference went pretty well and Erin's talk was well attended. At the last session as they were leaving from the farewell dinner the university dean asked Erin if she would mind moving to the island if he offered her a permanent position there.

Sam was there too and she couldn't help wondering if he had known this beforehand. He read the speculation in her eyes and leaned forward saying with a twinkle in his eye, "I told you that you were too good." He laughed throwing up his hands expressively, "And I thought I could keep you to myself!"

Erin gave him a quick punch in the ribs answering instinctively, "Well, from what I can see, you may well be stuck with me pretty soon, Dr. Sam Roy, whether you like it or not!" She turned and smiled politely at the dean and answered noncommittally, "I loved your island. Maybe I'll hear more about your proposal formally later on?"

Erin had insisted on spending the night at her hotel and the next day Sam picked her up from there for their planned glacier trip. He showed off his car to Erin. It was tiny, with a round top, manual gears and bright red. As Erin stared, he said with a touch of mock pride, "Don't you think it is as much fun as its owner?" Then eying Erin's genuine disbelief, he opened the passenger seat with a flourish, saying, "You can never tell till you try!"

They stopped at the seal resort before heading to the glacier. Erin giggled to see how much fun the dark and shiny seals were, splashing around playfully and how dandy they looked sunning themselves on the floating ice logs. The glacier park was as Erin had imagined—beautiful, remote and fascinating. She stood awed by the sheaths of ice cascading down the tall mountain faces, catching the sunlight and exuding glints of silver, steel and sharp blue, and then turned to Sam with a new respect in her eyes. It suddenly dawned on her that there was a part of him which was as grand and unique as those natural wonders in front of her. It was really not necessary to scale those dizzying heights to savor their power and magnificence. Just being there, having the privilege of seeing them, was a treat in itself. Erin was convinced Sam was happy here. He had loosened out a lot as she could tell from the way he talked and moved. His tension lines were gone and she was sure he was getting a lot of support in his experimental projects. Also the lifestyle here suited him for it seemed more relaxed and people actually took the time to enjoy the beautiful natural bounty all around. Sam had said he mostly biked or walked to work. Indeed he looked leaner and his tanned skin glowed. Yes, Erin was glad she had decided to come. Obviously Sam was thoroughly enjoying sightseeing with her. She felt good and wholesome. It was a simple emotion with no ifs or buts and she couldn't help feeling lucky that she had in her life someone who could actually make her feel this way.

Late that night before Sam came to see Erin off on her return flight he came to her hotel room. He took her hand and asked gently, "So are you going to marry me, Bonita? For that's why I had taken you to Hawaii last winter and that's what I was trying to propose to you that last night we were there."

Erin looked at him silently. She saw his earnest and anxious eyes and knew that however engrossed he may be with his work and studies she had always been on his mind just like he had been on hers. She gave a spontaneous laugh, "Really? Well, it seems to me you know my answer already, wise man of science, otherwise why did you stop me from saying yes, that last time?"

Sam said quietly, intently, "I just didn't want you to rush into anything Bonita. You had to make your decision on your terms and time. I had to be fair to you, right?" He paused, "I do have another poem if that's what it'll take to convince you again, but here nature all around is poetry, and that's so much more than just words." He stopped again and noticing she wasn't answering he continued, "I hope your answer will still be yes, Erin but even if you never ..."

This time it was Erin who put her hand on his lips and scolded him, smiling, "Fancy, a great scientist like you rushing to conclusions, Dr. Roy!" She hugged him impulsively, taking in the smell of cedar and sea that she had felt before. It was vital and unadulterated and Erin knew for sure it would safeguard her happiness and health for a long long time. Sam was still waiting, a pulse throbbed on his temple and his jawline was set and tense. So before moving away she quickly kissed him on the lips and continued "I repeat Dr. Sam Roy, the answer is yes. And I'll also be your first subject in whatever research project you want to do, whenever you want to do it." Erin paused and, reading the persistent question in his eyes, put her hand briefly on his heart and whispered deliberately "wherever you go."

"Are you sure, Bonita?" Sam breathed beaming. "For now, your life is totally in my hands." Sam laughed and, slipping his hand in his pocket, brought out the most delicately-crafted beautiful mother of pearl ring. It glinted and shone in deep jewel tones as it caught the light. "I'm sorry I don't have a diamond ring from the jewelry store for you. But I have to tell you, Bonita this is the first mother-of pearl that I had fished out from under water when I had gone on a scuba diving trip the last day we were in Hawaii last year. I had a local artist make this ring for you. I just guessed your size," he ended. Erin was so moved her eyes grew misty. "It's beautiful," she breathed in incredulous delight. And as Erin stood mesmerized and ecstatic at the sight of that elegant creation Sam slid the ring on Erin's ring finger and took her in his arms as if she were the most precious, most delicate and fragile treasure he had ever held. Erin felt oddly safe and complete even with her abnormal mutations.

Although she was leaving, Erin knew she would soon have to return for good to this wondrous man with his wondrous dreams on this wondrous island.

After Erin left, in between his increasingly busy social life, but with occasional tinge of sorrow about leaving Aurum University and missing Erin, Sam concentrated on recruiting many young scientists trained in microbiology, protein purification and characterization, as well as cancer biology, therapy and drug development. Because of his name recognition in the field, as well as a very attractive salary and somewhat different island life style, his advertisements for these positions in leading journals and newspapers received a high rate of response, particularly from Northern European countries, but essentially from all over the world. He was surprised that some of the best scientists from the US were eager to come. Sam knew that these tenured professors at esteemed universities and had no financial need to move. But in the cover letters they all uniformly reasoned that a country like US that led the world in biomedicine, technology, was making a U-turn in its approach to support science. Because of the financial mess that the country was in and the latest talk of sequestrations, there was just not enough funding to do cutting edge research any longer. New discoveries cost money. It took 3 billion dollars to sequence the 3 billion bases in human genome. Back then it was a questionable investment for the country. However, the race to sequence the human genome advanced biology, engineering, physics and mathematics in an unprecedented manner. New sequencing machines and technologies evolved, more physicists and mathematicians got involved in designing techniques and programs to snip together the sequence of the human genome. Tremendous numbers of new jobs were created, and a milestone achieved. Science knows no boundaries. Most practitioners of science have no prejudice to search for truth and to do what they believe in and to help mankind, they will go to any country that provides that opportunity.

So Sam even recruited some well-known established scientists who were both curious about this new institution in the middle of nowhere but also attracted to the fabulous facilities for both conducting basic research and clinical trials under careful scrutiny and supervision. They were aware of the development of unique genetic database amassed by enterprising companies in southern neighboring countries such as Iceland.

In about six months since his arrival at the BFL, a nickname given to the island country, Sam's search for new types of marine microorganisms isolated from the island's flora and fauna produced a plethora of new and interesting strains. Sam knew exactly what he was looking for—bacteria that would establish symbiotic relationships with many of the large coastal aquatic creatures susceptible to cancers, just as human beings, and would try to protect their hosts from invading enemies such as cancers. Sam's group was looking for proteins similar to rhabdosin that had both cancer therapeutic and preventive activity. Out of 365 independent isolates from many different animals, Sam's group found 15 bacterial strains that produced proteins with anticancer activity, about 10 of which also had strong anti-viral activity. Remarkably, three of these proteins also had entry specificity in cancer cells and cancer preventive activity as measured by inhibition of histological determination of pre-cancerous lesion formation in normal mouse mammary cells when such cells were exposed to a potent carcinogen such as dimethyl-benz-anthracene. Such inhibition of precancerous lesion formation varied from 70 to 80%, even at very low concentrations of these three proteins. One protein, however, stood out among the three. This protein not only exhibited all the characteristics mentioned above, but also exhibited a deep blue fluorescence, characteristic of many other marine organism-derived fluorescent proteins.

Sam was fascinated with the blue glow exhibited by the protein in solution. In his native language, Bengali, blue is called neel. So he named the protein neelazin. Given the preferential entry specificity of this protein to cancer cells to kill such cells by multiple modes of action, its strong cancer preventive activity and its intense blue fluorescence, Sam instantly knew that neelazin could also be a great diagnostic marker to locate tumors in the human body and could particularly be useful to detect metastatic cancers. He conducted extensive structural and functional studies with neelazin and noticed that neelazin had interesting structural and amino acid sequence similarity with a human protein known as a metastasis suppressor protein, pointing to neelazin's potential as a bacterial tumor metastasis suppressor as well.

The anticancer compound entry specificity in cancer cells, but not in normal cells, and cancer preventive activities with neelazin were all conducted in cancer cells grown in culture. The next challenge for Sam and his group was to find out if neelazin was actually effective as a therapeutic and cancer preventive agent without any major side effects, initially in animals and then in human beings. With his familiarity of rhabdosin's experimental trials in animals for its toxicity and efficacy studies, Sam knew what to do with neelazin for such studies. He had the world's best animal facilities and a well-trained group of young and enthusiastic researchers to conduct such studies, and the results they obtained were spectacular: no toxicity of neelazin in any animals including mice, cats, dogs and monkeys, even at very high concentrations, but significant therapeutic effects in animals with advanced tumors, leading to complete regression of such tumors in most cases.

It was particularly heartening to Sam's group, since they knew of the structural and functional similarity of neelazin with a human metastasis suppressor protein. Neelazin prevented metastasis of human breast and lung cancer cells when highly metastatic cell lines were injected under the skin of mice. Neelazin significantly reduced the number of metastatic foci in such mice. Sam's attention was next turned to conducting human clinical trials. The island nation had established a medical regulatory agency modeled after the European Medicine Agency to ensure that patients are treated with an experimental drug only after their informed consent, appropriate compensation, all possible safety measures and careful monitoring, and the ability to withdraw from the trial at any time.

Sam was perfectly agreeable to such arrangement but his recruitment of patients involved, among 20 others, Kathy, Nick and Anil. The three were sick, even though they were improving with rhabdosin treatment. Still, they successfully undertook the long and arduous journey to BFL for the trial. The initial trial was only to determine the toxicity of neelazin in such patients at various doses through i.v. injections, starting from a very low dose of 0.05 mg neelazin per kg of body weight through 5.0 mg per kg. The injections were given 3 times a week and the conditions of the patients, including imaging of the tumor volumes, were carefully monitored. Most of the patients, including Kathy, were at stage IV with multiple tumors. However, this initial Phase I trial for toxicity determination was very successful. Not only did most of the patients start to show enhanced appetite, significant weight gain and cheerful demeanors, but MRI, CT and PET scans as well as other imaging techniques showed significant regression of the tumors without any perceptible side effects.

Sam was particularly elated that both Kathy and Anil showed major shrinkages of their large tumors in just a couple of months, and by the time the trial terminated, they both started to recover substantially. Nick's recovery was slower, but he was cheerful, playful and walked around the beach, just as he did in his California home. Sam felt that it was a matter of time until Nick would recover completely, so long as he continued taking neelazin, even after the termination of the trial, under the watchful eyes of his physician oncologists. This required complex procedural matters and thoughtful decisions on the grounds of compassionate use from the BFL Medical Regulatory Agency. Sam was hopeful, however, that under the circumstances, and given the flexibility of decision-making in the island nation, that patients whose tumors were on the verge of being eliminated would be permitted to continue on the drug, even after the trial had ended.

One thing still bothered Sam. Long term i.v. injections were not only painful and damaging to the veins, but they occasionally led to inflammation and profuse bleeding through rupturing of the veins and surrounding blood vessels. He desperately needed an oral neelazin formulation, similar to an upcoming oral insulin for diabetic patients. But he was more upbeat on his previous conversations with Brian for expressing the bacterial neelazin gene from the human genome through gene therapy. Such an approach would not only eliminate i.v. injections or taking pills, but hopefully would also prevent cancer emergence.

Since neelazin's structure resembled that of human immunoglobulins and therefore it had no immunogenicity in his patients, Sam was hopeful that if Brian could figure out how to introduce the neelazin gene in the human genome, perhaps to allow its secretion in the blood as a blood protein where it could reach essentially all tumors, that would be the way to go. But it's such a long and un-travelled road! However, he had enormous confidence in Brian and decided to talk to Brian and update him on his latest findings on the efficacy of neelazin on tumor regression in his patients when given by the i.v. route. Brian's laboratory/clinic was next to Sam's building and Sam was eager to convince Brian to consider the gene therapy with Kathy, Nick and Anil, after they obtain appropriate clearances both from the Medical Regulatory Agency and of course from the patients, their relatives and guardians. They knew that they had to be patient, since such a decision would be taken, if at all, through prolonged discussions on the impacts of such a decision making process.

The proximity of Sam's lab and Brian's clinic was a blessing. Not only did they meet for occasional lunch in the cafeteria to discuss both social events and scientific progress, but they had combined lab meetings once a week and their students, interns and postdoctoral fellows worked and socialized freely in the tiny island. The ability of neelazin to allow shrinkage of the tumors in the 20 patients, which included Kathy, Anil and even Nick was a major triumph in their quest to treat cancer, and hopefully to prevent its recurrence. More important to this group was to find a way that would be painless, long term and would prevent cancer in vulnerable people with a family history of cancer who worry about when the cancer will strike them and where. This was the most frequent subject of discussion between Sam's and Brian's groups, and they ultimately agreed that introducing the neelazin gene in the genome of cancer patients, including vulnerable people, and observing them carefully over a long period of time for the total elimination of the cancer and its non-recurrence is the dream that brought them together in the island.

So Sam and Brian challenged the researchers and clinicians in their laboratories and clinics to figure out how to conduct such gene therapy trials in their patients, who seemed to be eager to go through it since they had great faith in Sam, Brian and particularly in neelazin as the dream anticancer drug. The BFL Medical Regulatory Agency also saw a major opportunity to draw the attention of the world to their home country as the center of innovative medical research and practice and issued a go ahead with this very limited gene therapy trial. Buoyed by this decision, both Sam and Brian focused their attention on the way to move forward.

On his next group meeting with his students and postdoctoral fellows, Sam was emphatic about how best to proceed with the gene therapy trials. He said to his group, "Unfortunately, one cannot go directly to human trials as one would have to show that genetic manipulations in mice will work fine and, more importantly, that neelazin does not have any ill effects on other tissues. The last thing we want to do is cure something but then have a bigger problem later. For example, people who had diabetes can be treated with a family of drugs that take care of their blood glucose problem. However, these drugs sometimes also make them obese. Talk about misfortune. Knowing that we have to be very careful with neelazin, I am thinking about introducing the neelazin gene in the genomic DNA of mice that are prone to developing breast cancer, and hope to demonstrate that our mice will not develop breast cancer and will not have any other problems."

"That's a great idea, but how will we do that?" asked a student.

"Let me tell you what I am thinking: We will take embryonic stem cells from breast cancer prone mice and introduce neelazin gene in their genome."

"But, you know, Sam, the neelazin gene may get integrated onto some essential housekeeping genes and we will never have any surviving mice!"

"I have an idea," suggested one of the students: There is a region in the mouse genome called the Rosa26 locus. Disruption of this region has no deleterious effect on mice: They are born, they eat, drink, fight, play, mate as adults and make offspring. So we will introduce the neelazin gene into the Rosa locus: That will be our genetic playing field and I guarantee we will win, and win big time."

"Can I have the naming rights for this?" asked one of the post-doctoral fellows. He offered, "We will call them "Rosazin cancer free mice."

"Of course."

One thing bothered Sam. Putting the neelazin gene in the chromosome is one thing, but neelazin protein must also come out of the cells to attack the tumors wherever they are. This question was bugging him: How do you bring the neelazin to the cancer cells?

Sam went to his office, made a cup of tea and turned the TV to the Bluefrostland basketball league game. After watching his team lose, he began flipping channels and found the documentary on World War II fascinating. It was a black and white film showing how fighter planes were dropping bombs. The images on the screen prompted an idea: "What if I make neelazin come out of blood cells and deposit it onto cancer cells?" He thought it might work. It was 2 am in the morning. He was tired but excited, so he called the lab with the expectation that at least one of the lab rats will still be working in the lab. Sam himself was a lab rat when he was a postdoctoral fellow—he worked day and night—and luckily he had a few lab rats in his own lab in Bluefrostland.

"Henry, this is Sam," he said.

"I know it's you, Sam," said Henry. "No one else calls at 2 am. What's up?"

"I have this idea that I would like to run by you and see what you think. I want to run an experiment."

"What is it?" asked Henry.

Sam replied, "As we discussed, I want to see if putting the neelazin gene in mouse hematopoietic stem cells (HSCs) will cure and prevent cancer. What I want to do is make a new form of neelazin that will be secreted from cells such that it can deliver the weapon onto cancer cells."

"Now that's an idea," said Henry. "Neelazin itself may not come out of the cells. I am sure it can be done, but let me read more about it and we can talk next day, I mean in the late morning when you come to the lab."

As a follow up to their conversation, Sam and Henry brainstormed, researched literature and found out that indeed they can insert a particular protein sequence termed secretory or signal sequence that allows protein to come out of cells without making big pores. They essentially sneak out of the cells. So Sam's lab generated DNA that contained this signal sequence along with neelazin sequence, and put it back into the HSCs of breast cancer prone mice. They performed a lot of analysis and indeed neelazin could now be found secreted into blood. The mice with reconstituted HSCs delivered neelazin to breast cancer cells and prevented them from growing. Now, Sam thought it really should work in humans. He wondered if he had not watched how air planes target bombs on enemies, he would not be able to generate this great idea about neelazin.

There was significant excitement at the lab meeting and the

medical students and postdoctoral fellows started the complex genetic manipulation of generating the breast cancer prone transgenic rosazin mice with the neelazin gene in their genomes. After months of failure, frustration and trying, the Sam Roy laboratory generated what they hoped for. Now the real test: Would they have cancer? The members of the laboratory followed the newborn mice day and night to make sure they do not have any visible health problems. Luckily for them, the rosazin mice looked healthy. More importantly, after a few months, while the non-rosazin mice developed breast cancer, the rosazin mice did not show any signs of breast cancer. Now that the biggest hurdles have been achieved using mouse models, Sam was convinced that it could be done with human hematopoietic stem cells. Sam said, "How can we extend these studies to show that it works in humans?"

The next day Sam came to the lab and gathered all his lab members. They were still celebrating the successful work of putting neelazin in mouse and preventing breast cancer. In the lab meeting, Sam started by saying that they would have to capitalize on the momentum in the lab. Once their paper would be published other laboratories would try to do investigations in humans. Since it takes about 6 months for a paper to get published and appear in print, Sam Roy's laboratory had a six-month head start. "OK, you all know that mice are not humans, but I believe we have the proof-of-principle that expressing neelazin from hematopoietic stem cells may hold the promise for curing breast cancer and who knows what other forms of cancer." He was excited that he could isolate bone marrow from the patients and using scientific protocol, his laboratory could isolate hematopoietic stem cells, the mother of all blood cells, and insert the secretable neelazin construct in the genome of such cells. They already had the neelazin gene cloned but the question is how to insert it properly in the human genome. What if this foreign gene inserts into a key housekeeping gene that keeps our cells and body going? Is it even possible to do this?"

So Sam went to a website called Pubmed, a scientific literature search engine for a online library that catalogs all published scientific discoveries. What Sam was looking for was whether there was any part of human genome around which he could insert the neelazin gene without affecting the cell function, similar to the Rosa locus in mice; a safe harbor to dock his neelazin gene. He was not confident that such a locus existed in human but, it was worth giving the search a try. A search for "safe harbor gene locus" surprisingly identified a paper published a few years earlier where docking of foreign genes did not have any adverse effect on cells or their viability.

Ok, now that they found the docking site or the destination, how is his gene going to reach that destination? Is there a GPS guide for human genome vehicle?

He was relieved when he searched for relevant research article. He pulled out the paper and read it thoroughly. "Oh my! Am I lucky or what? I was looking for only one thing but I ended up getting more than what I asked for." There were indeed papers that identified a particular region of human gene on chromosome 19 called AAVS1 locus where it was possible to dock a gene of interest without having any visible effects on cell growth. This reminded him of the game of treasure hunt: What is the next clue?"

The articles also described genetic engineering of a GPS guided molecular scissor called TALENs (Transcription activator like effector nucleases), very much like Zn-finger nuclease, that can make cuts at specific sites in genome. In their case it would be the AAVS1 locus. Once cut, the foreign gene containing the neelazin sequence could be inserted in a precise manner. This is exactly what he was looking for. What surprised him the most was that the original TALE factor was derived from a plant pathogen. Here he was ready to use plant and a marine bacterium derived factors to treat and hopefully prevent one of the deadliest forms of human diseases. Oh, how the bugs (bacteria) can bug the cancer!

The next set of experiments were done under the guidance of Brian Warten, who was now Sam Roy's friend and collaborator in the elegant laboratories at Bluefrostland.

Brian's laboratory isolated hematopoietic stem cells from patients and, using the TALEN technology, they engineered their modified blood stem cells. They spent a significant amount of time characterizing these cells. Indeed, these stem cells gave rise to all kind of blood cells and, importantly, they secreted neelazin in the culture media. Now they could put back these stem cells into patients after culturing them in laboratory petri dishes.

However, the biggest test was whether it will work in humans.

So, Brian wrote a clinical protocol in which he would enroll breast and other cancer patients that did not respond to any other treatments. The protocol was reviewed by his institution approval committee and approved. It was also approved by the Bluefrostland regulatory agency. Surprisingly, 20 patients from all over the world enrolled in this initial trial. All their expenses were paid for by the Bluefrostland authority. Brian put them in two categories: One group would receive their own unmodified hematopoietic cells while the other 10 would receive TALEN engineered neelazin hematopoietic stem cells. He fondly called them "Talezin." Of course the patients were informed of the trial but they did not know if they were receiving the unmodified or the engineered cells. The trial went smoothly and there were no adverse side effects on the patients, except for some pain and initial fever. In the control group, the cancers kept growing. However, in a few months, 7 of the 10 patients of the group that received engineered cells showed significant signs of tumor shrinkage, while 2 did less so, and in one case there was no change in tumor volume. The outcome of the trial was so clear-cut it was stopped and the control group was administered Talezin cells. This was one of the most triumphant human clinical trials. To this day 15 of those 20 patients are living virtually free of cancer.

Once successfully done in 20 patients, Brian and Sam combined their forces to use the same strategy in Kathy, Anil and Nick, who arrived in Bluefrostland a month ago. All three had different types of cancer. Kathy had ovarian cancer, Anil had colon cancer and Nick had a muscle tumor called rhabdomyosarcoma on his leg. To their utter delight, and to some extent to their expectation, all three of them had their symptoms greatly reduced. In six months they recovered without any visible trace of the tumors.

twenty one
Sam's award ceremony

The banquet hall in the hotel in Bluefrostland was aplomb with dignitaries, colorful lights and beautifully adorned circular dinner tables that had the name plates of all the attendees of the evening. There were dozens of newsmen and TV reporters, including several from the US, who were eagerly waiting for the event to kick off. The mayor, the consulate generals of half a dozen countries, famous oncologists, chairman of the Medical Council of Bluefrostland and many others had already arrived. In short, these were the "who's who" of the island city and academia.

After the initial perfunctory introductions, the keynote speaker of the evening made quite a dent on the audience, to an extent that people almost looked at him in unison. As if there was something great happening: "And now, ladies and gentlemen, let me have the pleasure of introducing you to the medical genius of the decade, who has successfully employed bacterial protein to completely cure cancer patients, the one and only, Dr. Sam Roy."

There were just about thirty steps to the stage and then four stairs to reach the podium from where Sam was seated. Yet, these thirty steps presented themselves like 30 eventful milestones in his life. At every step, Sam had a different picture in his mind. He cherished each step, because each of them brought in collages of events from his life. His mother's smiling face. He imagined how she would have beamed today in this audience. His medical senses combatted with his emotions … for this one second, he wished there was something like his mother's soul that was witnessing this moment.

Then there was the S-Class Mercedes Benz crashing against the Banyan tree. Something that was perhaps ordained for him. Something that undeniably steered him into this research.

He had walked three steps already.

Then there was Anil-da. Smiling and intelligent, simple and plain in lifestyle yet the one person who infused lofty ideals in him. The guy who bombarded his life with practical experiments on science, like burning a paper with a magnifying glass, or making a kaleidoscope with broken pieces of colorful bangles and three strips of broken rectangular glasses, when their home was being renovated. The pinhole camera, the electric radio that they made together when he was 16.

Another step and another memory revisited.

There was Sonali's face … and then there was Erin's face. The face of the lady for whom he had already started to have a comfortable feeling. Is it love? He was still unclear. He did not ask questions to his mind. He just let it flow.

There was his father, sweating, yet not forgetting to bring his exercise book while returning home, or the almost molten chocolate for his little sister.

There was this little room, the little balcony, the extremely small living room. An agglomeration of every little square inch, all of which screamed for an identity, the square inches that made up a 640 square foot little home, yet replete and resplendent with a lovely bonded family. A home, that reeked of phenyl disinfectant in the single bathroom, turmeric and onions in the kitchen and incense sticks every evening in all the rooms, of which were but three. One living cum dining space and two bedrooms, one of which was converted into a study room for the two children. The little fights that he had with his sister. But then, being the big brother, he was taught to make the sacrifices. Like giving a bigger piece of cake to her, or letting her use his nice comfy chair once in a while for studies.

Three more steps …

There were many other flashing pictures. Pictures of his favorite teachers and also teachers that insulted Sam. Dr. Ganguli's proud face. The next door neighbor's gentle old father who always wished him well. There were so many thoughts popping up like bubbles but immediately bursting at the onset of the next bubble.

He had reached the podium.

The whole world was listening live to what he had to say. There were cameras from TV channels, newspaper reporters and so many other channels that were eagerly waiting to gulp down his sound bytes. Was he nervous? No. But there was something that choked him slightly. For a moment he looked at the glass and then took a sip of plain water that was kept covered on the podium.

And then what happened was not quite under his control. He stammered a bit …

"I … a. I … err. Uh … ahem. He cleared his throat. Anil-da's smile came and faded away.

He felt more confident now.

"I am used to speaking about technologies, medical processes and treating patients. I am indeed honored to be a part of this discovery. But I would also acknowledge all the people, the gentle patients and the kind-hearted people in my life who have held my hands and walked me to this juncture. The ward boy, the professors of the university, the secretarial staff, the adjunct professors, the janitors, the lab assistants and everyone who directly or indirectly contributed to a peaceful research environment, I thank all of you from the bottom of my heart. I miss many people very dear to me, whom I could not save, because this discovery was not conceived at that time. But I fervently hope that millions of mothers will not lose their children to this dreaded disease, millions of sons will not lose their parents before time and millions of dear relatives will not hang pictures of their loved ones on the walls of their homes 10, 20, 30 or 40 years before their due dates, before they would normally leave this earth!"

There was loud applause from every corner of the room. Sam went on "I also thank the little mice, creatures of God, my true friends and martyrs, who we took for granted throughout the experiments and who did not have not the slightest idea as to what a big giant step to mankind they have helped provide. I thank all the brave, kind patients who, with their unflinching faith in me, leant their bodies for research. I also thank their families for allowing me to test on them for the betterment of humankind.

Let us not become complacent. And, as a great philosopher and poet once said '… miles to go before I sleep.' We have the daunting task of successfully proving all our findings on continuing clinical trials and then move on to next phase of drug development. There are various parties to be looped in our journey and we all move as a team and not as individuals. The Bluefrostland regulatory authority, the patent authorities, judicial system, drug companies, all have joined hands. We fervently look forward to their help and support in helping us bring our vision to fruition. Else this will only be another thesis paper that will remain in the obscurity in someone's hard disk drive. Thank you and God bless you all.

Above all, I would like to thank the common man for not losing hope as soon as they hear the six letter word C-A-N-C-E-R invading their families or friends. This is a disease that we have already started to stare in the eye, and we are going to take it head-on till it withers away. I would earnestly request all of you to have faith in me and my team of researchers. It is a marathon race and we are in the 20th mile and there is no turning back now. We HAVE TO MAKE IT TO THE FINISH LINE. Thank you all and God bless you."

Whew. "God bless you," coming from him? Sam was himself surprised.

Sam raised a toast saying, "To a cancer-free world."

Sam received a standing ovation, and the camera operators literally fought to have a closer shot of him.

While he walked down the stairs from the stage to the ground, he felt his eyes moisten a bit. If he missed anyone badly at this moment, it was his mother, no one else. Was it his mother who prompted those last words, "God bless you?" He had always been an atheist and had spent hours debating against God, only to irk his mother and make her raise her voice.

twenty two
Celebration dinner

B lue Moon Lookout was a quaint little restaurant on the northern coastal area of Bluefrostland. It was set on a hilly ledge in the midst of undulating fields of the prestigious Heritage Golf Club. Because of its elevated position the back windows had a gorgeous view of the coast as well. Both Sam and Brian had taken many a happy swing with their clubs on the unspoiled acres surrounding this neat eatery. However, recently, work pressures extending to after-hours had kept them away. So, after the success of their somatic gene therapy trials on Kathy, Nick, Anil and others, Sam and Brian decided they deserved to celebrate by having dinner together at this favorite joint.

Six months had gone by since the trials had ended. The neelazin gene had been successfully inserted into the DNA of all the patients, with all of them doing well. Sam took particular interest in his three old patients, Kathy, Nick and Anil. Their tumors had shrunk. Anil could walk again. He was getting ready to go back to India. Nick had been tutored when he was at the hospital. Soon he would go back to school. Kathy was feeling perfectly normal. She had asked her boss in San Francisco if she could return to work.

Since this was such a special occasion, Sam and Brian had also invited their respective heads of departments and their lawyer friend, George Rivera, to dinner at this fancy restaurant in Bluefrostland. George arrived from the US about a week ago, and was marveling at the scenario.

Sam and Brian arrived early. They were sitting in the outdoor garden, sipping drinks when they saw Mary Thompson and George Rivera entering together and talking with the hostess. Brian went over to talk with them.

Brian approached George. "George, I am really glad that you could make it to the dinner. Let me introduce Dr. Mary Thomson, head of oncology at my university. Mary and I go a long way back. Mary, George is our patent lawyer. He has helped file a patent application for Sam and me in the US before our somatic gene therapy trials here. We have some future plans for germ line therapy and George is going to help us."

Mary extended her hand. "Glad to meet you, George. Brian, did you have a good trip in from your home?"

As they were talking, Sam spotted Andrew Timis, the head of microbiology at the University of Bluefrostland, and they both came over and introduced themselves to Mary. George looked around appreciatively at the surrounding green, observing, "This is a really nice place you guys picked for dinner, so lush. What a great celebration of neelazin expressed from genomes, with the intent of eventually vanquishing cancer. If someone on the street were to hear about this, they would think you were making the whole thing up."

Brian nodded in appreciation. "Glad you like this place. Nature is indeed a treasure house of wonders if we know where to look. Basic research is kind of like that. If you look at the stem cell technology and what induced pluripotent cells can do to replace damaged cells, particularly in neurologically-affected patients or people with damaged vertebra in car accidents, it's just amazing. I think somatic cell gene therapy, even using a foreign bacterial gene rather than a functional human gene, will achieve even greater success than stem cell therapy."

George nodded. "I agree, which reminds me of the patent controversy involving stem cells in Europe. Mary and Andrew, are you interested in hearing about it?"

Mary nodded yes.

Sam interjected, "If I may interrupt, please. I already booked a table by one of the back windows, so we can talk in peace and also enjoy the coastal night lights. Since we are all here, why don't we all get our drinks from the bar and go inside?" With the group's consent, the party moved inside.

Inside the Blue Moon Lookout, the walls were hung with country artifacts and a lovely painting of a crescent blue moon in a dark starry sky. Noticing Mary admiring the painting, the owner of the restaurant came up and explained that when the water mists roll in from the sea often times the moon looks blue violet through the haze giving this place its name. The group made their way to the back, to the table Sam had reserved.

After taking a seat, Mary addressed the table. "As a surgeon, I am focused on my patients getting well. Unfortunately, many of them don't. I only know about the importance of patents superficially. Our legal department colleagues are always talking about patenting. I wish they would explain it further. George, can you elaborate on the importance of patenting?"

"Sure," said George. "As far as I am concerned, it is important for taking a concept to a final product, particularly in the cases of medical products and processes. Mary, how much time does your department spend securing a substantial grant?"

"Oh, I don't even want to think about it!" Mary shook her head.

George replied, "However, if this new process is patented, you can raise a ton of money by licensing. Since my first year of law school I was fascinated by intellectual property and the US patent laws. As a matter of fact, US commerce owes much to the patent laws. I think the patent laws in Bluefrostland may catch up eventually. It's too early to tell. Do you know that the patent laws are in the constitution written in 1790?"

"1790?" Andrew looked up from his menu. "Isn't that soon after the independence in 1776 when the forefathers got together to write up the Constitution?"

George excitedly put his glass down, waving his hand to emphasize his point.

"Exactly! Our forefathers were visionaries, but to be truthful, also had some vested interests in having patent laws on the books. Jefferson, himself an innovator, believed that there would be an explosion of inventions and that commerce would flourish in our country. The first US patent was granted on July 31, 1790."

Sam looked puzzled. "I did not know that the history of the patent law dates that far back."

"It gets even more interesting," said George. "Do you know who signed these patents?"

"Well, the patents were signed by the director of the US Patent & Trademark Office (USPTO)," offered Sam. "Who was the director of patents in the 1790s?"

"Aha!" George leaned into the table, gesticulating with his index finger. "Not only the first patent, but several patents issued subsequently were signed by President George Washington and Secretary of State Thomas Jefferson as well as the Attorney General. Of course, the number of patents granted at that time was lot fewer than the ones granted now. Imagine our current president signing the issued patents, even using an electronic signature."

"Now I see what made the United States so great and prosperous," Sam said, turning with a quick grin to Mary. "The President of the United States signing the granted patents shows high level of government support."

George offered, "This is just one part. When George Washington read the third patent issued on automated flouring mills, because of his interest in automation, his mill and several other mills took up the patented process by paying fees to the patent holder."

Mary was interested. "George, now that you mention it, I remember reading in a biography of Abe Lincoln that he held a patent. During one of his official addresses, he actually stated that the patent system was important for the United States, to promote innovations and to reward innovators of new technologies. I don't remember his exact words, but I think he said something like 'The patent system stood to prevent copying and encouraged the development of innovative technology by providing fuel to the fire of genius.'"

Sam reclined in his chair, "I had no idea that the patent system and the institution of USPTO were so important. But George, you said something about the European patent laws that bar stem cell patenting in Europe, which is allowed in the US. Why so?"

"I'll tell you in a minute, but shouldn't we order some dinner first?" George asked. "I mean I personally think that this restaurant has a fine cellar and I don't mind talking as long as my glass gets refilled, but you may all be starving. What do you say?"

"Of course, of course" said Sam, and Brian gestured to the waiter to please take their orders.

Once the waiter left, George was back on track, "This has something to do with the European patent laws that do not allow patenting of anything that is against public order or morality. The recent EU court decisions consider human embryonic stem cell patenting, or patenting anything involving higher forms of life or humans, as unacceptable and not allowed under European patent laws. They do allow patenting of microorganisms though."

Sam offered, "Again, I have read about a US Supreme Court decision, I think around the early eighties, that basically said 'anything under the sun that is made by man' is patent eligible in the United States. That sure sent a strong message to the technology developers in the US."

"You are absolutely right!" George fisted triumphantly. "Those justices certainly knew about the intention and vision of our forefathers who put the patent laws in our constitution. As far as I remember, Jefferson specifically urged the US Government that ingenuity should get liberal encouragement from the government to help promote new ideas and technology. Look at the stem cell patent situation in Europe and in the US and you can begin to appreciate what the USPTO stands for."

Mary sipped from her wine glass, but her head was cocked in attentiveness to the conversation. "Give us a significant example of a court case," she said.

George responded, "Well, there was a revolutionary court case about stem cells in the mid-nineties. Several groups demonstrated that mouse embryonic stem cells could be converted to mouse heart cells in the presence of a combination of growth factors. The credit for generating the first stem cell lines from human embryos goes to Dr. James Thomson who subsequently obtained US patents on the procedures for obtaining cell lines that could differentiate into a variety of human tissues in presence of appropriate cues. This process could then be used to repair damaged tissues in human patients leading to a field in medicine called regenerative medicine."

"I think Dr. Thomson is a distant relative of mine," said Mary.

"Oh?" said George. "Then perhaps you will appreciate what I am talking about. Although there were initially some questions regarding prior art, meaning public disclosure of the procedure for obtaining the stem cells, the patents were granted in the United States. In contrast, a former colleague of Dr. Thomson, Dr. Oliver Brustle of Bonn, Germany, submitted a patent application in Germany on procedures for using embryonic stem cells for generating neural precursor cells. In 1990 Germany had enacted the Embryo Protection Act which forbids research on human embryos, but did it not refer to the embryonic stem cell lines which were then unknown. Dr. Brustle was granted a patent in 1999 for his procedure to derive neural precursors from human embryonic stem cells. These embryos were leftovers from fertility clinics. Since such leftover embryos are normally discarded, Dr. Brustle argued that using such embryos to derive neural precursor cells with the potential application in the treatment of brain disorders was a better use."

"There was a challenge, wasn't there?" asked Sam.

"Yes," replied George. "In 2005, after Dr. Brustle received his patent, the non-profit organization Greenpeace challenged the issuance of the patent in Europe invoking the 'order public' clause in the EU Directive on the Legal Protection of Biotechnology Inventions that applies to all EU countries. Such a directive describes the use of human embryos for commercial purposes as against public morality and therefore illegal. Initially, a judge in the Federal Patent Court in Munich ruled in favor of Greenpeace. On appeal, the Appeals Court also sustained this ruling, whereupon Dr. Brustle took the case to Germany's Supreme Court. In November 2009, the Supreme Court deferred the case to the European Court of Justice for a final resolution which could address the patentability of human embryonic stem cells, since such cells result from the destruction of human embryos. In October 2011, a 13 judge panel of the European Court of Justice upheld the Greenpeace argument that it is immoral under the EU laws to patent a procedure that requires destruction of human embryos and therefore patenting of human embryonic stem cells. This ruling from the European Court of Justice cannot be appealed further and is applicable to all members of the EU.

While they were still munching on the last pieces of fried calamari, artichoke pate, and steamed shrimp in lettuce wraps, their dinner arrived steamy and inviting. Mary had ordered a vegan "steak," which was comprised of a patty of wild rice, beans and chopped green and red peppers along with a mixed green salad. Sam had an herb and almond crusted fillet of salmon with rice pilaf, and, for health reasons, Brian had selected a lightly poached chicken breast with boiled veggies. The others had steak that was prepared sizzling and rare at their tables, with everyone marveling as the flames from the portable grill crackled and shot, red and orange, high into the air.

As they settled down to their plates and appreciatively dug in, Andrew raised a thought. "I have a question, George! You said stem cells and many animals have been patented in the US. Are you saying that we can even patent Brian and Sam's creation of a new Kathy, Anil or Nick because their genomes have changed forever with a bacterial gene inserted in them?"

George finished his bite, transferring his napkin to his lap. "Now you are getting into complex but very interesting issues involving patents. Patents are meant to exclude people from copying your invention for profit for a limited period of time. Obviously, you cannot patent human beings, but you can try to patent the process of introducing a foreign gene in the human genome. Are you aware of a patent application filed by a couple of people in New York to patent a hybrid human-animal?"

"Yes," said Mary. "I have heard about it. I thought it was sick."

"Certainly," George agreed, "it was not the most politically correct patent application. The USPTO rejected the application, I think around 1998 or 1999, because the 13th amendment of the US Constitution forbids ownership of slaves or for that matter any human being, including hybrid human-animals."

"What is this hybrid human-animal?" asked Andrew curiously.

"It was never actually created," said George, "but these people in New York took advantage of goat-sheep hybrids called Geep. The DNA sequence similarity of a goat and a sheep is more or less of the same order as that of humans and chimpanzees. So the patent applicants argued that it should be possible to make hybrids of human and chimpanzees, following the example of the Geep. They wanted to patent them to bring attention of people to the potential threats of altering human genomes for all kind of purposes such as organ harvesting, better and stronger animals for hard labor, etc., provided the hybrid human-animal was not considered a human. As I said, the PTO rejected the application based on the 13th amendment that prohibits ownership of a human being, even though it was not clear how much genomic identity or alteration a human genome must have to make it a non-human for patenting purposes."

Brian had finished his chicken and was starting, tentatively, in on his veggies. "Are you saying that even though we cannot patent Kathy, Anil and Nick, we can perhaps patent the process of genetically altering a human being by describing and claiming the technique of introducing a foreign gene like the neelazin gene in the human genome?"

"Yes, certainly!" said George. "I have never quite understood the PTO's position that ownership of an animal-human hybrid is forbidden under the 13th amendment. Patenting is not ownership. The purpose of a patent is to prevent other people from making commercial or financial gains for a limited period of time, usually 20 years from the date of filing of the patent application. So how does the 13th amendment come into play?"

Mary replied, "George, let's not push the issue. I like the USPTO's position on rejecting the patent on hybrid human-animals. Let's talk about the feasibility of filing a patent application on the process itself rather than the genetically-altered humans such as Kathy."

"I am all for it," said George, "because I know that if I file a claim on the human, it has essentially no chance of success of ever getting a patent even though it's a great innovation. Even if the PTO accepts it, I know that social and moral advocacy groups will sue in a court of law and a judge or a panel of judges will rule against the patent granting because many judges do not view science from the perspective of scientists."

Brian, in between mouthfuls, offered, "I don't know about you folks, but I like George's good work. I have great confidence in George in making the right claims in our patent application. George, do you think we should be able to patent human gene mutations? And I mean mutations, not the isolated and purified human genes that the Supreme Court found non-patentable in the Myriad Genetics court case. Given the great utility of BRCA1/BRCA2 or even other human gene mutations, shouldn't they be patentable?

Sam interrupted the conversation. He was keenly aware of Erin and Kathy's family history of breast and ovarian cancer and the importance of diagnosing such mutations that predispose them to the cancer. Will future screening of such mutations be developed without patent protection? He was also aware of some of the behavioral problems being attributed to potential mutations in the human genome. He continued, "Some of these societal issues are also complex. For example, those related to complex behavioral problems such as depression, proneness to violence, etc., can be studied scientifically. You can make mutations in female mice where the mutation will lead the mother mouse to completely ignore its kids. It remains a controversial discovery as some scientists do not agree with the phenotype. Are such mutations present in human mothers and fathers who are accused of child abuse and go to jail? Indeed, if some child abuse cases are shown to be due to mutations, rather than personal attitudes, shouldn't such mothers be sent to medical clinics rather than to the jail? I also read about a case where three generations of a family, grandfather, father and his child were found guilty of rape and murder and committed in death row in Philadelphia. I wonder if these people might inherit some kind of a mutation, similar to BRCA1/BRCA2 mutations that confer susceptibility to cancer, but leads to criminal inclinations?" said Sam. "Of course, it might be difficult to separate the environmental factors such as parental nurturing, lack of decent housing or education, poverty etc., from actual genetic factors. Unlike BRCA1/BRCA2 gene testing, such investigations may lead to stigmatization and stereotyping of a whole group of people."

Brian raised his head from his almost empty dish and said, "I definitely think that something should be done now. Let me ask if anyone wants dessert. I, for one, enjoy their crème brûlée', which reminds me of my mom's home-cooked caramel custard but I also recommend their excellent cheese cake with wild berries. Anyone game?"

Mary ordered mixed berries but all the others except Sam had the cheesecake. The strong black coffee felt good after their drinks. Andrew said after a long pause. "I have another question. If Sam is right that his bacteria produce neelazin-like weapons to keep enemies, such as cancer, off their habitat, what about other illnesses that can also kill their hosts? I am thinking more like heart attack, diabetes etc. Can neelazin or perhaps other proteins be effective as a weapon to treat these illnesses? Also, can other bacteria with a long term residence in the body produce similar but different weapons, to give us a whole new arsenal of weapons acting as new drugs?"

Sam replied, "Good question. All I know is that there is a company in India that used similar rationale to isolate a protein, different from neelazin, from another bacterium with anticancer and anti-HIV/AIDS activities but I don't know any details about it. Besides, if you express neelazin from the genome of people with cardiac problems and/or diabetes, you can quickly find out if the expressed neelazin in the blood of these peoples will help in ameliorating their diseases."

"Sam," Brian replied, "It has been a long day and night. I had too much alcohol as well. At least, I am not driving. I will take a taxi to my hotel."

George said, "Wait a minute! Before we go home, Brian and Sam should decide on filing a patent on the neelazin expression method as soon as possible. Since US patent law recently changed to a first-to-file method, your intellectual property should be protected by filing as soon as possible. Word may get out for such an important invention. Let us file a patent on the somatic gene therapy portion of your invention first here in Bluefrostland. We can then use the Patent Co-operation Treaty to file patent applications in the United States, Europe and other offices."

"You mentioned something about the issues with advocacy groups?" said Sam. "Won't they object?"

"I have been looking into that matter, Sam," replied George.

In the country of "free will," if an individual willingly accepts the treatment to cure themselves from this deadly disease or use it as prevention, religious organizations and advocacy groups may not have very strong objections. Even courts of law will likely rule for you. I cannot promise that you will not be harassed by the media, but think of the large number of people who will be thankful to both of you."

Brian added, "As you explained last time we met, you will have some funds to continue with our studies and find more cures. At least I hope so! Thank God we are in Bluefrostland where we do not have to worry too much about funding. We would rather do research."

"Well said, Brian," Andrew nodded.

George added, "I have done some preliminary market research on the royalties you can charge for treating a single person. The general estimate is about $5K for treating a single person."

"Thanks for the estimate," said Sam, "but I believe the university has staff members who can help us with the royalty pricing."

"Certainly," said George. "That is something of a rough idea that I am using to make my point. One more thing—Sam and Brian mentioned that this treatment can be modified and applied so that the children of people with the likelihood of getting cancer will not have cancer. Did you think about the possibilities of germ line therapy?"

Brian agreed, "Yes, it is quite possible that we can come up with such a treatment."

The friendly waitress arrived with the dessert dishes and the hostess, who happened to be the owner's wife, carried a large pot of freshly brewed coffee. The aromas blended to give the dining hall a pleasant ambiance.

Eying his sizable cheesecake slice fondly, George said, "Patenting germ line gene therapy is a complicated beast. Some of you probably heard about the Monsanto's Roundup-Ready Soybean controversy. Mr. Bowman, a 75 year-old Indiana farmer was accused of infringing on Monsanto's patent on soybean, because he obtained seeds from a grain elevator and planted them without paying a royalty to Monsanto. Monsanto's patent covers not only the original herbicide-resistance gene but the patent is supposed to cover self-replicating future seeds that will arise out of the original soybean. These seeds will, however, still harbor the patented genes and therefore will be covered by the patent. The district court, the Court of Appeals for the Federal Circuit (CAFC), and now the Supreme Court, ruled for Monsanto's claims."

"I get it!" exclaimed Sam. "You are saying that treating a human to protect her progeny by germ line therapy may be similar to the Monsanto's self-replicating soybean seeds since they will harbor the patented neelazin gene in their genomes for generations and therefore will be covered by our patent claims." He gulped down the last morsel of his crème brûlée.

"That's right!" chimed George. "It is sort of strange because this is about treatment for a deadly disease. Shouldn't a mother have a say on the long-term protection of her child?"

Andrew replied, "Let's not think that far ahead, and deal with the present issues instead. I am sure that Sam and Brian will get in touch with George soon for filing the patent applications for somatic cell gene therapy. I am sure that George will put a couple of claims on germ line therapy as well even though we are not going to try germ line therapy at this time."

Sam shifted in his seat. "I have one final question for George, since he mentioned a royalty payment of $5 thousand for any clinician who will use our patented somatic cell gene therapy. Isn't that quite a bit of money for a clinician to pay just for using a procedure? I read in the newspapers that part of the legal problem Myriad Genetics faced in enforcing its patents on BRCA1 and BRCA2 genes and their mutations were the alleged high fees they charged for the mutation screenings, and look what happened! The Supreme Court has now declared that isolation and purification of genes such as BRCA1 and BRCA2 separated from the chromosomes are not patentable inventions because they are basically the same genes that occur in nature. I am shocked at this legal reasoning, even though I agree with the conclusion. I think isolated and purified BRCA1 and BRCA2 genes should be patent eligible if they have great utility, which they don't. The mutations are, of course, a different matter, which is the reason why many women go to get themselves checked by the Myriad Genetics. Aren't patented antibiotics the same as those produced by the microorganisms in nature? The only reason they are allowed patents is because they are isolated and purified and have great utility in combating infections. So, nobody complains. Will the Supreme Court now invalidate all the patents on antibiotics? Then, there are human genes and all bacterial genes that do not have what are called introns in them and thus should be patentable under the new verdict of the US Supreme Court? So how one gene can be patented but the other cannot be creates a lot of puzzle and confusion? I am sure the law will be modified in near future as new cases start to arrive in the CAFC and the Supreme Court. Of course I am expecting our patent application will raise several issues: First can you patent a newly created somatic cell that is personalized? Second, if you cannot patent the cells, can you patent the procedures for generating the cells, and last but not the least, should altering human genome for one generation by inserting a bacterial gene be patent eligible? I think with this patent application we will really be pushing the boundaries of patenting in the context of 21st century.

Mary pushed her chair back and stood up. Her cheeks looked as rosy as the summer berries still strewn on her plate. "I think I'll have to leave now, but it has been such a …" she faltered as Sam stood up.

Sam looked at her, empathetically. "As you said, your main interest is to cure patients, so I hope all the other legal jargon didn't exhaust you too much."

Mary smiled, her eyes twinkling. "What I enjoyed was how we all brought medicine, law and, of course, science of microbiology onto the same table. Quite a daunting, no, let me correct that, quite a daring feat, gentlemen!"

As she turned back to the group a final time, Mary offered, "Mark my words, this is just the beginning!"

The rest of the group, looking at their watches, realized how quickly time had passed. Shaking hands, and convivial as they made their way outside, they realized that the evening's discussions were indications of many more to come.

twenty three
Life begins

On Mother's Day, after Sam, Brian and others returned from Bluefrostland, Sam and Erin were married. They had a simple wedding at a local Buddhist temple in Milanburg. This was Dr. Lee's temple and he was thrilled to give Erin away. Only close friends and relatives were invited. There was no religious ceremony, but a passage from the *Gita* was read by Preeti, Erin's friend. Erin's aunt Angela read a hymn from the bible. Brian accompanied her on the piano. Kathy and Nick flew in from California to attend the wedding. Nick was the "best boy." In spite of fervent requests from Sam, Anil could not stay. He had already returned to Kolkata. He sent his blessings via Skype. Sam and Erin lit candles and bowed down together in front of the smiling pictures of their mothers, imagining that the mothers were showering them with blessings from heaven. Nick, who had no memories of his mother, read a poem that he had written for his mother, as Sam and Erin knelt.

A simple dinner was provided by Sam's friend and neighbor, the owner of the coffee shop. As Sam and Erin snuggled together on their first night as husband and wife, Erin spoke softly, "Sam, I am scared."

"Why, sweetheart?" Sam looked deeply into Erin's eyes, smiling. "This should be the happiest day of our lives."

"I am scared of the white elephant in the room," Erin offered, drawing her face away from Sam. "We paid homage to our departed mothers today. But they haven't really left us. We have their genes. What is going to happen to our children, Sam? What if we have a daughter and she inherits a double dose of defective genes from us. How can we prevent her from having cancer? Isn't she doomed?"

Sam looked at Erin with kind eyes. "I see that along with your heart, your brain has gotten soft. I hope it is only a temporary effect, just the wedding blues. First of all, you will have to get somatic gene therapy like Kathy and you will be making neelazin yourself to nip your tumors in the bud. Then, the DNA of our baby can be altered using *in vitro* fertilization and the baby will be equipped with neelazin, too. Lastly, it is just a matter of time, before we come up with an effective germ line therapy for adults. We will be able to fix you up so that all our kids will be able to attack cancer cells. Of course, we have to jump through legal hoops and wait for permission, but it is bound to happen. Now, I will hear no more of this nonsense about cancer. Enjoy your new husband, and that's an order."

Sam shut her off with a passionate kiss and turned the light off. There would be plenty of days ahead to discuss such matters. For the moment, Erin was rendered speechless.

Epilogue

On a cool but balmy April morning in Chicago, five years after they all returned from Bluefrostland, Erin woke up with a telephone call from Kathy. Kathy's name was all over the news media because of the publication of a patent application covering her recovery from ovarian cancer, because of the somatic cell neelazin gene transfer to her genome. She recalled vividly how, within a couple of months after Brian and Sam gave her repeated injections of some cells in the hospital in Bluefrostland, her pain and discomfort subsided, she could walk around the house and had less depression and anxiety than ever before. Within a year, when she returned to the US and started to take long walks in the neighborhood parks and streets, she felt good about herself with her spirits up, and even started telling jokes. Brian and Sam visited her in her home. That was almost a year ago. Both Brian and Sam were in good spirits and in a jovial mood. They showed Kathy some pictures that they described as MRIs or something like that, and some animal stuff like CAT or PET scans. There were no pictures of animals though. Instead, what these two gregarious people explained to her were the sizes of the tumors in her ovary. They did shrink, and almost disappeared!

After a couple of months, when Kathy knew she was feeling great, with her anxiety and depression levels essentially gone, the tumors were gone too. Apparently, all these results were described in the patent application which just came out from a published patent application from the USPTO, and the reporters from all over the country had a field day not only with the inventors but also with Kathy, since either Brian or Sam had to divulge to the reporters who the experimental patient was. Although Nick's situation was also described in the published patent application, and his grandparents were contacted, he had a reprieve because of his age. Kathy, along with Sam and Brian, however, became instant celebrities because of the exploratory nature, potential future application and the apparent success of the new way for cancer therapy.

Erin was delighted. She was aware of most of the results from her occasional discussion with Sam, but was warned to keep the on-going results confidential. With the publication of the patent application, now all was out in the open and she did not have to bear the burden of keeping the news secret any longer. Kathy and Erin spoke for hours on the telephone, and started to contemplate a celebratory party with all their friends and relatives, including of course Brian and Sam, as well as Nick and his grandparents, if they were available.

Before the party date, however, other events started to cast a shadow on their dream party. News started to appear in many newspapers and magazines that activist organizations, particularly the Society for the Ethical and Moral Preservation of Human Genome Integrity were contemplating a court challenge of any patent likely to be issued by the USPTO or any patent offices of other countries. Many religious organizations also began to openly support such a proposition. However, many relatives of cancer patients also started a movement for positively supporting the patenting of the somatic cell gene therapy involving a cure for cancer. They argued that without patent protection, such treatment modalities are not likely to move forward, even with substantial government support, and therefore patents need to be issued to bring the clinical research to the bedside and the marketplace. They further emphasized that this was the intention of the US Congress when they passed the Bayh-Dole Act in 1980 to allow academic institutions and small businesses ownership of any patent arising out of Federal Government funding. Isn't such a law designed to promote patent protection of new inventions in academia, as happened in Bluefrostland?

The USPTO had no official response to the publication of the patent, but after a year and half issued the patent with Brian and Sam as co-inventors. The official explanation given was that the patent application met all the statutory requirements such as novelty, non-obviousness, great utility and enablement, with a detailed description of the experimental procedures. Many supporters of the patent indignantly pointed out that the US is not Europe, that moral and public order clauses are absent in the US patent laws, and indeed starting from Thomas Jefferson in 1790 to the Supreme Court decision in 1980, US patent laws encourage that ingenuity should get liberal encouragement from the US Government and that anything under the sun that is made by man is patent eligible in the US. Thus, the stage was set for a legal resolution of a patent that claims permanently modifying the human genome with a foreign bacterial gene, which everybody hoped will guard against not only cancer but other invaders of the human body such as viruses, parasites or perhaps even deadly diseases such as heart attacks!

Within a month after the patent was issued, a group called Society to Defend the Sanctity of Human Genome brought a lawsuit against the USPTO, with Brian and Sam as defendants in the District Court in Chicago. The society argued that the patent covers the ownership of human genome which is God's and nature's gift to mankind and should not be a market commodity or the property of profit making people. After almost a year, a judge in the District Court revoked the patent, citing the Supreme Court decision on patenting of human genes, although in this instance the gene was the bacterial neelazin inserted in the human genome.

After their glorious dinner conversations more than five years ago when George was given the go-ahead to file the patent application, the parties met again in a restaurant for dinner. The atmosphere this time was very different. Nobody was smiling or making jokes or talking about the differences in patent laws in different countries. They learned that unlike in science where progress is rapid and past experimental results seldom count, in the law, old precedents matter. But they had an option. They could appeal the District Court judge's ruling to the CAFC to reverse it.

George, Brian, and Sam reflected over their plates.

George looked at Brian and Sam. "I am amazed at the District Court decision. Do you guys have any thought about what we should do now?"

Brian fiddled with his fork. "What options do we have?"

George replied, "We could appeal to the CAFC."

Sam, pragmatic, quipped, "How much will it cost and how long will it take?"

"The time frame is hard to say," said George, "but it will probably take a year or more. It could be decided by a 3-judge panel, or depending upon the public outcry or other exceptional circumstances, it could be decided *en banc*, meaning that all the judges may choose to participate in the decision making process. The cost will depend upon how many or what types of litigating lawyers we hire, but will likely cost not less than half a million dollars."

"Half a million dollars?" smirked Sam. "My university will never come up with this kind of money to defend our patent."

"Well," said Brian, "ever since this controversy started and the District Court ruling was issued, I have been getting many letters from many cancer patients and their relatives, particularly patients with stage III or IV cancers. They all want to get this gene therapy, which they hope will give them a new lease on their lives. Many offered financial help should we need it. There are many lawyers whose relatives are cancer-stricken and they offered to help. So I think money will not be a major problem."

George nodded, feeding off of Brian's hope. "Actually I also got a number of letters from many lawyers who even offered to argue the case. So I also believe that we can defend our patent at CAFC." In the ensuing conversation, Sam and Brian assented.

George, with his attorney friends, filed an appeal to the CAFC to reverse the District Court ruling. Many institutions and even individuals with interest in the outcome of the case, submitted *amicus curiae* briefs to the CAFC, either in support or against the issuance and validity of the patent. Brian and Sam also received a lot of hate mail and angry phone calls, but were comforted that they also received many supportive letters from people who thought that the patent would help bring new therapies for cancer and other deadly diseases, not only from the introduction of the neelazin gene in the genome but perhaps many similar bacterial genes whose products would be non-toxic, non-immunogenic but with significant efficacy to fight cancer and other diseases.

The diversity of strong opinions over gene therapy and the lack of any precedent made the oral arguments not only highly emotional but also full of legal precedents, or a lack thereof, as well as lively discussion on whether it should be patentable. Many quotes referenced the wisdom of the European Union legal system which forbids patenting of anything involving humans or higher forms of life, including human embryonic stem cells. Lawyers for Brian and Sam pointed out that the US patent laws have no such clause; the constitution urges the US Government to provide liberal encouragement to ingenuity and the ofted-quoted Supreme Court decision that "anything under the sun that is made by man is patent eligible in the United States, including life forms of all kinds." The lawyers defending the patent also pointed out that the patent only covers the process of introducing a foreign gene in the human genome to fight and guard against deadly diseases, not the human beings. After about a year of intense debate, the CAFC ruled for granting the process patent, much to the delight of Brian, Sam, and a host of their supporters. The decision was, however, a split 2-1 decision. The majority cited a vision inspired by Thomas Jefferson's purview, reflected in the 1980 Supreme Court decision that anything with human intervention may be patented under US patent laws as long as it meets the statutory requirements of patentability and shows potential utility for public good.

This 2-1 majority CAFC decision was not well received by the losing party. Their belief that human beings, including their genomes, are sacred and precious gifts of God and nature, were too preciously guarded to let the CAFC decision stand. They filed an appeal to the US Supreme Court to review the CAFC ruling and turn it down. Initially, the Supreme Court granted certiorari and accepted the case, but later refused to consider it with the recommendation that the subject requires a new law and the intervention of the US Congress. Congress formed a subcommittee to consider not only the patent eligibility of the process of modifying human genome but also a doctrine of fairness to neighbors whereby the US Government was urged to build a fence both in the Southern and Northern borders to keep illegal immigration in check.

Sometime, during their stay in Bluefrostland, Erin had undergone gene therapy along with other patients of Brian. Like all the other volunteers, she, too, was hoping to live a life free of cancer. Both Sam and Erin also hoped to be parents some day. They expected to raise their children in Milanburg, but they would continue to visit and work in Bluefrostland every summer.

As the world around them fought battles of jargon and policies, Sam and Erin held on to the dreams they had dared to share in the faith and hope that one day there would be a way. The colored sun of Hawaii and the cold blue sun of Bluefrostland looked so different, but hadn't they brought the same light and meaning to their lives?

References

General references

The emperor of all maladies: a Biography of Cancer. Siddhartha Mukherjee (2010)

Emerging Cancer Therapy: Microbial Approaches and Biotechnological Tools. (2010) A. M. Fialho and A.M. Chakrabarty, John Wiley & Sons, Hoboken, NJ.

The Immortal Life of Henrietta Lacks. Rebecca Skloot (2010)

Article on the theory of bacteria evolving as mammalian mitochondria

Redox proteins in mammalian cell death: An evolutionarily conserved function in mitochondria and prokaryotes. V. Punj,. and A. M. Chakrabarty. Cellular Microbiology 5: 225-231. 2003

Articles on azurin and p28 as anticancer agents

Bacterial proteins: A new class of cancer therapeutics. A.M. Chakrabarty, *Journal of Commercial Biotechnology* 18: 4-10. 2012

Patent controversies and court cases: Cancer diagnosis, therapy and prevention. A.M. Fialho and A.M.Chakrabarty *Cancer Biology & Therapy* 13: 1229-1234.2012

A first-in-class, first-in-human, phase I trial of p28, a non-HDM2-mediated peptide inhibitor of p53 ubiquitination in patients with advanced solid tumours. .M.A.Warso, J.M., Richards, ., D. Mehta, D., K. Christov, C., Schaeffer, C., L.R.Bressler.,T. Yamada, D. Majumdar, S.A. Kennedy, C.W. Beattie and T.K.Das Gupta, *British Journal of Cancer* 108: 1061-1070, 2013

References about gene therapy

Genetic Engineering of human pluripotent cells using TALE nucleases. *Nature Biotechnology*, 2011: Vol. 29, 731-734. Dirk Hockemeyer, Haoyi Wang, Samira Kiani,Christine S Lai, Qing Gao, John P Cassady, Gregory J Cost,Lei Zhang, Yolanda Santiago, Jeffrey C Miller, Bryan Zeitler, Jennifer M Cherone, Xiangdong Meng, Sarah J Hinkley, Edward J Rebar, Philip D Gregory, Fyodor D Urnov, Rudolf Jaenisch.

Functional Genomics, proteomics, and regulatory DNA analysis in isogenic settings using zinc finger nuclease driven transgenesis into a safe harbor locus in the human genome. *Genome Research*, 2010: Vol. 20, 1133-1142. Russell C. DeKelver, Vivian M. Choi, Erica A. Moehle, David E. Paschon, Dirk Hockemeyer, Sebastiaan H. Meijsing, Yasemin Sancak, Xiaoxia Cui, Eveline J. Steine, Jeffrey C. Miller, Phillip Tam, Victor V. Bartsevich, Xiangdong Meng, Igor Rupniewski, Sunita M. Gopalan, Helena C.Sun, Kathleen J. Pitz, Jeremy M. Rock, Lei Zhang, Gregory D. Davis, Edward J. Rebar, Iain M. Cheeseman, Keith R. Yamamoto, David M. Sabatini, Rudolf Jaenisch, Philip D. Gregory, Fyodor D. Urnov.

Hematopoietic Stem Cell Gene Therapy with a Lentiviral Vector in X-Linked Adrenoleukodystrophy. *Science*, 2009: Vol. 326, 818-823. Nathalie Cartier, Salima Hacein-Bey-Abina, Cynthia C. Bartholomae, Gabor Veres, Manfred Schmidt, Ina Kutschera, Michel Vidaud, Ulrich Abel, Liliane Dal-Cortivo, Laure Caccavelli, Nizar Mahlaoui, Véronique Kiermer, Denice Mittelstaedt, Céline Bellesme, Najiba Lahlou, François Lefrère, Stéphane Blanche, Muriel Audit, Emmanuel Payen, Philippe Leboulch, Bruno l'Homme, Pierre Bougnères, Christof Von Kalle, Alain Fischer, Marina Cavazzana-Calvo, Patrick Aubourg.

Gene therapy for Leukodystrophies. *Human Molecular Genetics*, 2011: Vol. 20, R42–R53. Alessandra Biffi, Patrick Aubourg, and Nathalie Cartier.

Age-dependent effects of RPE65 gene therapy for Leber's congenital amaurosis: a phase 1 dose-escalation trial. *The Lancet*, 2009: Vol. 374, 1597 – 1605. Albert M Maguire, Katherine A High, Prof Alberto Auricchio, J Fraser Wright, Eric A Pierce, Francesco Testa, Federico Mingozzi, Jeannette L Bennicelli, Gui-shuang Ying, Settimio Rossi, Prof Ann Fulton, Kathleen A Marshall, Sandro Banfi, Daniel C Chung, Jessica IW Morgan, Bernd Hauck, Olga Zelenaia, Xiaosong Zhu, Leslie Raffini, Frauke Coppieters, Elfride De Baere, Kenneth S Shindler, Prof Nicholas J Volpe, Enrico M Surace, Carmela Acerra, Arkady Lyubarsky, T Michael Redmond, Edwin Stone, Junwei Sun, Jennifer Wellman McDonnell, Bart P Leroy, Francesca Simonelli, Prof Jean Bennett.

Gene therapy trial for Haemophilia. *British Journal of Haematology,* 2008: 140, 479–487. Samuel L. Murphy and Katherine A. High.

A mouse model for X-linked Adrenoleukodystrophy. *Proc. Natl. Acad. Sci. USA,* 1997: Vol. 94, 9366–9371. Jyh-Feng Lu, Ann M. Lawler, Paul A. Watkins, James M. Powers, Ann B. Moser, Hugo W. Moser, and Kirby D. Smith.

Adenovirus vector-mediated gene transfer in hemophilia B. *N Engl J Med.* 2011: Vol. 365:2357-2365. Nathwani AC, Tuddenham EG, Rangarajan S, Rosales C, McIntosh J, Linch DC, Chowdary P, Riddell A, Pie AJ, Harrington C, O'Beirne J, Smith K, Pasi J, Glader B, Rustagi P, Ng CY, Kay MA, Zhou J, Spence Y, Morton CL, Allay J, Coleman J, Sleep S, Cunningham JM, Srivastava D, Basner-Tschakarjan E, Mingozzi F, High KA, Gray JT, Reiss UM, Nienhuis AW, Davidoff AM.)

References about the Supreme Court decision on the Myriad Genetics patent case

Patenting human genes and mutations: A personal perspective. A.M. Chakrabarty, *Journal of Commercial Biotechnology.* 19: 3-5, 2013

Glossary

-da: A suffix added to the proper name of a man in the state of Bengal, India, to show respect, say Anil as Anil-da. Generally used to address people older than the subject.

Adrenoleukodystrophy (ALD): A severe and rare genetic disease, meaning linked to X-chromosome. In ALD, beta oxidation of fatty acids is impaired which results in the accumulation of very-long chain fatty acids, mainly in central nervous system.

Azurin: A protein produced by the bacteria *Pseudomonas aeruginosa* shown to have anticancer properties in mice.

Germline gene therapy: Replacement of genes in germline cells. The changes in the DNA of germline cells are propagated in the progeny/offspring.

Gully: a small street, usually unpaved.

Hematopoietic stem cells: These are the progenitor cells that give rise to all blood cells.

Microbiota: All the microorganisms found in a certain organ.

Mitochondria: A structure in animal cells that provides energy.

Neelazin: A fictitious protein with blue fluorescence. Neel means blue in Sanskrit and Indian languages like Bengali.

Namak Haram: A person who betrays you after being fed by you, after eating your salt. Namak means salt.

Peptide: A very small protein, or part of a larger protein.

Retrovirus: A virus that uses RNA as its genetic material.

Rhabdosin: A fictitious protein from a fictitious bacterial species called *Rhabdosis pulmoneria*.

Somatic cell gene therapy: Replacement of a defective gene with a functional gene in somatic cells. Somatic gene therapy does not propagate into next generations.

TALEN: An abbreviated version of Transcription Activator-Like Effector Nuclease, an artificial designer enzyme that cleaves a very specific and desired region in the gene or genome.

Thali: Food served on a metallic plate and several metallic bowls.

Zn-finger nuclease: A designer enzyme, similar to TALEN, able to cut specific DNA sequences.

Related Titles from Logos Press®
http://www.logos-press.com

Building Biotechnology
Scientists know science; businesspeople know business.
This book explains both.
Yali Friedman

Hardcover ISBN: 978-1-934899-29-8
Softcover ISBN: 978-1-934899-28-1

Great Patents
Advanced Strategies for Innovative Growth
Companies
David Orange, Editor

Softcover: 978-1-934899-18-2
Hardcover: 978-1-934899-17-5

Biotechnology Entrepreneurship
From Science to Solutions
Michael Salgaller, Editor

Hardcover ISBN: 978-1-934899-13-7
Softcover ISBN: 978-1-934899-14-4